UNDERSTANDING HUMAN BIOLOGY

LABORATORY EXERCISES
SECOND EDITION

Mimi Bres • Arnold Weisshaar
PRINCE GEORGE'S COMMUNITY COLLEGE

Benjamin Cummings

San Francisco • Boston • New York
Cape Town • Hong Kong • London • Madrid • Mexico City
Montreal • Munich • Paris • Singapore • Sydney • Tokyo • Toronto

Executive Editor: Gary Carlson
Project Manager: Crissy Dudonis
Executive Managing Editor: Kathleen Schiaparelli
Assistant Managing Editor: Karen Bosch Petrov
Production Editor: Patty Donovan
Supplement Cover Manager: Paul Gourhan
Supplement Cover Designer: Victoria Colotta
Manufacturing Buyer: Ilene Kahn
Associate Director of Operations: Alexis Heydt-Long
Cover Photograph: © Shutterstock/Dmitry Kosterev

ISBN-10: 0-13-179009-9

ISBN-13: 978-0-13-179009-4

Benjamin Cummings
is an imprint of

18 2023

www.pearsonhighered.com

Contents

	Preface	iv
Exercise 1	Introduction to the Scientific Method	1
Exercise 2	Using the Compound and Dissecting Microscopes	23
Exercise 3	Functions and Properties of Cells	43
Exercise 4	Investigating Cellular Respiration	65
Exercise 5	Enzyme Activity	83
Exercise 6	Food Analysis and Choices for Good Health	97
Exercise 7	The Skin: Example of an Organ	115
Exercise 8	The Musculoskeletal System	135
Exercise 9	Examination of Skeletal Structure	155
Exercise 10	The Nervous System	175
Exercise 11	Introduction to Forensic Biology	191
Exercise 12	The Circulatory System	211
Exercise 13	Introduction to Anatomy: Dissecting the Fetal Pig	231
Exercise 14	Organ Systems in the Abdominal Cavity	251
Exercise 15	The Reproductive System	275
Exercise 16	Mitosis and Asexual Reproduction	297
Exercise 17	Connecting Meiosis and Genetics	313
Exercise 18	Human Genetics	349
Exercise 19	Introduction to Molecular Genetics	369
Exercise 20	Biotechnology: DNA Analysis	393
Exercise 21	Using Biotechnology to Assess Ecosystem Damage	421
Appendix	Self Test Answers	437

Preface

Understanding Human Biology: Laboratory Exercises is designed for a one-semester, non-majors, general biology course with a human focus. The topics and exercises are general enough to be compatible with most introductory-level human and general biology texts currently in use. The activities demonstrate that basic biological concepts can be applied to a wide variety of plants, animals, and microorganisms.

This book is unique not because of the specific topics covered but because of the approach to these topics. The laboratory exercises are planned to help your students

- gain practical experience that will help them understand lecture concepts
- acquire the basic knowledge needed to make informed decisions about biological questions that arise in everyday life
- develop the problem-solving skills that will lead to success in school and in a competitive job market
- learn to work effectively and productively as a member of a team

The basis of scientific work is asking questions and answering them by observations or experiments. Thus, we feel strongly that the most important goal of an introductory course in human biology should not be to simply outline facts about anatomy, psysiology, or new techniques. Rather, students should come away with some understanding of the processes of investigation that are basic to science and how scientists work to solve problems.

We hope that working through this laboratory manual will be an exciting experience for both students and instructors, and one that will leave students better prepared to meet the demands of our increasingly scientific society.

SPECIAL FEATURES FOR YOU, THE INSTRUCTOR

Three-Pronged Approach to Laboratory Learning

- **Elicit Interest.** Students perform better when they have a clear understanding of the reasons and objectives behind each exercise. Students are often more willing to proceed if they know why they are asked to do each activity and what objectives they are expected to accomplish during the laboratory period.
- **Clear Directions.** Students need clearly and concisely written directions on how to proceed. This is the first and last science class for many students and we can make very few assumptions about the lab skills they might bring into the class.

- **Establish Relevance.** A well-designed lab needs to help students understand the meaning and importance of the findings they just gathered, and to establish relevance between laboratory and lecture topics, and with everyday life.

Clarity

- All content and problem-solving goals are accomplished by using a **simple, non-threatening approach** to the lessons. Even students with little science background will be able to **get involved** and master the material.

Instructional Objectives

- A clear set of **Instructional Objectives** is provided at the beginning of each exercise. This helps students identify the most important concepts that will be emphasized in each activity.

Focus on Thinking and Problem Solving

- One of the most important features of this laboratory manual is the **infusion of thinking and problem-solving skills into the content areas.** Research has shown that it is easiest to master thinking and problem-solving skills when thinking and content are learned together. **Understanding Human Biology** is based on this approach. Activities and questions within an exercise build on previously learned information and encourage students to transfer information from one section of the course to another.

Team Building Opportunities

- Most laboratory activities emphasize a **team approach.** Group work is encouraged and often required. In the real world, we are expected to interact with others to solve problems and complete projects. This approach provides opportunities for students to work together, share ideas, and function effectively in groups to accomplish tasks.

Real-life Connections

- Questions throughout the activities and those incorporated into the **Self Tests** are designed to connect the laboratory activities to timely subjects and real-life experiences. The exercises are intended to stimulate interest in topics that can help them make decisions regarding their own health and nutrition, understand current topics in the news, and their role in the environment.

Many Opportunities to Practice Skills and Reinforce Concepts

- Activities are **reinforced with relevant questions,** many more than can be found in any competing editions. It is not only the number of questions but also the **type** of questions that are important. Questions are designed to provide **continuous feedback** and **elicit targeted thinking.** This keeps the learner "on track" and fosters development of problem-solving skills.

- Through **questions interwoven** into each activity, **Comprehension Checks,** and **Self Tests,** recall of important concepts is reinforced. Students also have the opportunity to transfer content knowledge gained to new situations.

These Lessons Work!

- These laboratory activities have been **fully tested** by thousands of students on this and other campuses around the United States.
- Experience has shown that the simplified structure and easy-to-understand language of this manual works well with a diverse student population.

Requires Minimal Supplies and Equipment

- These exercises are cost-effective and affordable even for schools with budget constraints. **Most schools will already have all the equipment needed to perform the lab activities.**

Designed for Simplicity

- All activities are **easy to set up and maintain.** Our program is large and multi-sectional. Despite large numbers of instructors and students, this laboratory curriculum is incorporated smoothly in all sections. The simplicity of the laboratory approach has allowed new personnel to be **"up and running" with minimal preparation time.**
- Since the exercises require minimal equipment, many activities throughout the manual are **suitable for use in online sections.**

Adaptable to Any Curriculum

- Each exercise is designed for use in a three-hour laboratory period, but can easily be adapted to accommodate two-hour or 90-minute sessions. To provide **maximum flexibility** for instructors, each exercise is **divided into activities.** Activities can be deleted or presented as demonstrations without diminishing the value of the remaining components.
- Laboratory exercises can be presented **in sequence or rearranged** to suit your needs. The manual includes 21 exercises, allowing you the latitude to tailor the laboratory curriculum to your course and the equipment available at your institution.

Support Materials — Instructor's Guide

The Instructor's Guide provides you with

- complete preparation and setup instructions for each activity
- a complete Answer Key for all activity questions
- helpful hints to facilitate smoothly running laboratories
- master charts for recording results of classroom experiments

SPECIAL FEATURES FOR YOU, THE STUDENT

Informal Style

- The text is written in an active voice with easy-to-understand language. Key terms and important definitions are highlighted with bold print for easy recognition. Large spaces are provided throughout the manual for you to record and fully explain your answers.

Active Learning Experience

- Every exercise gives you an opportunity to be an active participant in the scientific method. You will form hypotheses, set up experiments, collect data, record your data in graphs and charts, and draw conclusions from your experimental results.

Self-guided Approach

- The laboratory procedures are clearly outlined to allow you to progress through each activity independently.

Tools for Success

The following components of each exercise will help you to succeed:

- **Instructional Objectives.** The objectives are listed first in each exercise so that you will be able to focus your attention on the main concepts of each activity.
- **Content Focus.** Each exercise includes a brief discussion of the background information that you will need to understand the subject of the exercise and prepare you to complete the activities that follow.
- **Notes and Cautions.** Note boxes provide helpful hints for solving problems and accomplishing laboratory tasks. Pay special attention to caution boxes that provide important safety information.
- **Comprehension Checks.** Stop and complete the Comprehension Check questions to get immediate feedback on your understanding of the basic principles covered in each activity. These questions also provide a chance for you to apply what you have learned to situations outside the classroom.
- **Check-off Boxes.** These boxes allow your instructor to check your progress before you move on to a new activity.
- **Self Tests.** Answer these questions after completing the laboratory exercise. Self Test questions allow you to assess your comprehension and apply your knowledge. Answers to the Self Test questions can be found in the Appendix at the back of the book.

ACKNOWLEDGMENTS

A project of this size and scope can't be completed without support and cooperation from many people. We would like to express our deep appreciation for their contributions.

Thanks to

Louis Renaud, Chairman of the Department of Biological Sciences at Prince George's Community College, for providing the opportunity to do this project and supporting professional development.

Michael Garvey, Biological Technology Center, Washington, D.C., for providing your time and technical expertise in the area of forensic molecular genetics.

Janet McMillen, Professor of Biology, Prince George's Community College, for reviewing newly developed laboratory activities. Her helpful suggestions contributed substantially to the effectiveness of these laboratory exercises.

Kenneth Thomulka, Assistant Professor of Biology, University of the Sciences in Philadelphia, for allowing us to adapt an imaginative and unique application of biotechnology to the educational arena.

Charlotte Weisshaar, for proofreading and editing our many manuscripts, thus ensuring continuity within and between exercises.

Robert Ewing and Meg Caldwell Ryan, for the excellent artwork and photographs they contributed to this manual.

The biology students of Prince George's Community College for classroom-testing our laboratory exercises and offering suggestions for improvements.

Andrew Sobel, Associate Editor, Biology, Prentice Hall, for lending your skill and support to this project. Your efficient and effective management played a significant role in making this second edition possible.

The reviewers listed below graciously gave their time to comment on the manuscript and suggested many useful changes. For their help, many thanks.

REVIEWERS

John S. Campbell, *Northwest College*

Renee E. Carleton DVM, *Berry College*

Alessandro Catenazzi, *Florida International University*

Clare Hays, *Metropolitan State College of Denver*

Dwight O. Kamback, *Lake City Community College*

Will Kleinelp, *Middlesex County College*

H. Roberta Koepfer, *Queens College, CUNY*

Monica McGee, *University of North Carolina–Wilmington*

William Simcik, *Tomball College*

Jason Yoder, *Itasca Community College*

EXER_CISE

1

Introduction to the Scientific Method

Objectives

After completing this exercise, you should be able to:

- use the scientific method to solve problems
- organize information to facilitate analysis of your data
- draw graphs that present data clearly and accurately
- interpret data in tables, charts, and graphs
- draw conclusions that are supported by experimental data
- analyze data using common statistical measures
- apply your knowledge of the scientific method to real-life situations

CONTENT FOCUS

What is science? What do scientists *do* all day? For most of you, these aren't easy questions to answer. The widespread picture of a middle-aged man in a white lab coat doesn't apply to most scientists. So, what are scientists really like? They all have the **"three Cs"** in common.

Just like you, scientists are **curious** about the world around them. They ask questions about everything. Can my diet cause heart disease? Why does the river look brown instead of blue? How can squirrels remember where they bury their nuts? Why do some cars get better mileage than others? Science is a method for answering these and many other questions.

Scientists don't accept things without **collecting information.** All the facts relating to a problem or question have to be carefully explored and checked for accuracy.

Scientists are **comfortable with new concepts.** If a better explanation can be found, scientists aren't afraid to give up old ideas for new ones.

To make the three Cs happen, scientists have developed a series of steps in investigation called **the scientific method.** Through trial and error, the scientific method has proved to be an efficient and effective way of attacking a problem. You've probably used some version of the scientific method many times in your life — without being aware of the steps you were following.

Note:

All the information, concepts, and relationships you'll read about in your textbook and laboratory manual were discovered and verified by the same scientific process you'll use in this exercise.

During this course, we'll be summarizing hundreds of years of study and experimentation. Information presented on television, in newspapers, and on the Internet often hasn't been confirmed by this same careful process and may not be correct.

ACTIVITY 1

FORMING HYPOTHESES TO SOLVE PROBLEMS

There are several ways that a problem can come to your attention. Someone may **assign** you the problem (this happens often in a school or a work situation), the problem may **thrust itself** upon you (your car won't start), or you may discover the problem by simply being **curious** about something you've seen. An easy way to attack the problem is to make an **educated guess** about the possible solution to the problem. It's an "educated" guess because you use all the background information that's available when making your guess. In scientific terms, an educated guess is called a **hypothesis.**

Let's begin with a simple situation that you might face in your college education.

The Problem

You were absent from chemistry class the day your professor gave out instructions to mix the chemicals needed for your laboratory experiment. Your roommate was in class and copied the instructions for you. You rush off to chemistry lab and mix the chemicals, but when you use them in your experiment, they don't perform as expected.

In **Table 1-1, list three hypotheses** about why the formula didn't work. Don't forget — hypotheses must be **testable!**

TABLE 1-1
CHEMISTRY EXPERIMENT HYPOTHESES
A: If the formula did'int work its because the your roomate copied down the wrong farmula and you need to talk to your teacher
B: You mixed the formula wrong.
C: Didn't have adiquet equement

Check your hypotheses with your instructor before you continue.

Some hypotheses can be tested by observation only, but, more often, you'll need a **combination of observation and experimentation** to be sure about the accuracy of your results. To understand how scientists work, you must follow the steps of the scientific method as they are used in actual **experiments**. In Activities 2 and 3, you'll **see how scientific method skills** are used to **set up experiments** and analyze the **information (data)** that's collected.

ACTIVITY 2   TESTING HYPOTHESES

The Problem
Investigate the effects of fertilizer on plant growth.

STEP 1:

You form a hypothesis about what you think will happen.

Hypothesis: Adding fertilizer will make plants grow taller.

STEP 2:

You design an experiment that compares the growth (in height) of plants that **receive fertilizer** with those grown **without fertilizer. Your design might be similar to the following:**

Begin with **20 plants** of the same type and approximately the same size.

■ An experiment is designed to isolate the factor you're interested in testing. All **other conditions must be held constant.** In this way, you're sure that your **observed results** were caused by the only factor that was varied.

■ Since you're investigating the **effect of fertilizer**, you'll want to hold all **other factors constant** (plant type, plant size, pot size, amount of water, amount of light, etc.) to avoid confusion. This way you can be **sure** that any differences in height are **due to the presence of fertilizer** and not due to some other factor.

It's helpful to plan an experiment with a **group** of plants (or animals). There are **two good reasons** to use groups:

■ If **unexpected factors** (such as death, disease, or accidents) affect a few experimental subjects, it won't ruin the experiment.

■ **Natural genetic differences between individuals of the same species** will cause some plants to grow taller than others (just as some people grow taller than others). You can separate this effect from that of the fertilizer by measuring the height in a **group** of plants for each treatment (fertilizer and no fertilizer). Because there's so much variation among individuals, you can increase the validity and reliability of your results by testing **as large a group as possible.**

STEP 3:

You decide that **ten plants will receive identical, measured amounts of fertilizer** each week. These are the **experimental** plants. They are receiving the treatment (fertilizer) that will help you test your original hypothesis (fertilizer will make plants grow taller).

Ten other plants will receive no fertilizer. These are the **control** plants. They don't receive the experimental treatment. You'll use these for **comparison with the experimental group** to help you interpret your results and to show that any observed differences in height between the two groups are due to the **only difference between them** — the application of fertilizer.

In the above example, there are **ten replications of the experimental treatment and ten replications of the control treatment.**

You plan to measure the growth of your plants (height in centimeters) **once a week for a month.** You'll keep detailed **records** of your **observations.**

✓ Comprehension Check

1. Why is it necessary to divide the plants into two groups (a control group and an experimental group)?

* unexpeted factors

* Natural genetic differences between indiuduals of the same species.

* To compare results.

2. Why is it important to keep conditions exactly the same in the control and experimental groups, **except** for the application of fertilizer?

* To make sure that your hypothis is right

* To see the diffrince between the two

* So no other factors come in to play besides your fertilizer

3. If you were designing this experiment for a fertilizer manufacturer, what changes would you make in the experimental design to improve the reliability of the results? **Explain** your answer.

* Increase the group size. 100 fertiized plants vs 100 non fertilized plants.

* Try differnt type of plents. this way you can see what happens to other plents.

* use differnt amounts of fertilicer.

Check your answers with your instructor before you continue.

ACTIVITY 3   INTERPRETING DATA

The month is up. You're ready to draw conclusions from your **data** (the information you have recorded). You'll be thinking about what your results mean and whether your hypothesis is **supported.** The information you collected during your experiment is presented in **Tables 1-2 and 1-3.**

✓ Comprehension Check

1. Were there differences in growth between the control and experimental plants?

    If so, which group grew taller?

    experimental plants

2. Do the results support the original hypothesis? **Explain** your answer.

    Yes, fertilizer made the plants

3. Why is it more accurate to compare the **average** height gain of the control and experimental groups (instead of comparing individual plants)?

    Takes out factors that you could'nt account for.
    If you have a bad plant or one that just takes
    off.

Check your answers with your instructor before you continue.

TABLE 1-2
HEIGHT GAIN (cm) OVER FOUR WEEKS — CONTROL PLANTS

PLANT NUMBER	INITIAL HEIGHT (cm)	WEEK 1	WEEK 2	WEEK 3	WEEK 4	GROWTH OVER FOUR WEEKS (cm)
1	10.0	1.6	2.0	3.0	2.5	9.1
2	11.5	2.2	1.5	1.5	2.0	7.2
3	9.6	1.5	2.3	2.6	2.0	8.4
4	9.2	2.0	3.0	2.8	1.5	9.3
5	10.2	2.3	1.2	1.6	2.0	7.1
6	11.0	3.2	1.7	2.0	3.2	10.1
7	10.0	2.6	3.0	3.0	1.4	10.0
8	9.7	4.0	2.6	4.0	2.3	12.9
9	10.4	DIED	—	—	—	—
10	10.4	2.3	2.3	2.7	2.7	10.0
TOTAL HEIGHT GAIN — 84.1 cm						
AVERAGE HEIGHT GAIN — 9.3 cm						

TABLE 1-3
HEIGHT GAIN (cm) OVER FOUR WEEKS — EXPERIMENTAL PLANTS

PLANT NUMBER	INITIAL HEIGHT (cm)	WEEK 1	WEEK 2	WEEK 3	WEEK 4	GROWTH OVER FOUR WEEKS (cm)
1	9.6	4.2	5.0	3.0	4.7	16.9
2	9.8	6.0	4.0	5.5	5.0	20.5
3	10.3	5.3	5.5	3.6	4.2	18.6
4	11.0	2.1	3.2	6.2	3.8	15.3
5	10.1	3.4	4.0	4.4	4.0	15.8
6	9.2	4.7	3.1	3.1	4.0	14.9
7	9.5	4.2	5.2	3.9	3.6	16.9
8	10.0	3.3	6.0	5.6	4.2	19.1
9	9.7	5.8	6.1	6.5	5.0	23.4
10	10.4	5.1	3.4	5.8	5.3	19.6
TOTAL HEIGHT GAIN — 181.0 cm						
AVERAGE HEIGHT GAIN — 18.1 cm						

ACTIVITY 4

PERFORMING AN EXPERIMENT AND COLLECTING DATA

Regardless of where you live, you've probably heard a lot of news stories lately about the increasing spread of HIV (which causes AIDS) and other sexually transmitted infections throughout the United States. In this activity, you'll simulate how multiple sexual encounters by an infected person can lead to the spread of a disease through a population.

1. On your laboratory table, you'll find a **dropper bottle of distilled water and a dropper bottle of phenolphthalein solution** (a chemical indicator).

   From the supply area, get **four clean sampling cups and a paper cup that contains an unidentified liquid.**

> ### Note:
> The unidentified liquid in your cup represents the body fluids that may be exchanged during sexual encounters. Your paper cup contains either distilled water or a solution that changes color when exposed to phenolphthalein. A color change simulates body fluids that are infected with HIV. Only one cup in the class is "infected."

2. Read through the following instructions on how to perform "body fluid" exchanges.

   WHEN YOUR INSTRUCTOR TELLS YOU TO START, **follow the steps listed** and complete **Exchange #1.**

   To "exchange" body fluids, follow these steps:
   - Randomly select **one partner** for the exchange.
   - Pour **all** of the solution in your cup into your partner's cup.
   - Your partner will pour **all** of the solution in his/her cup **back into your cup.**
   - Pour *only half* of the solution back into your partner's empty cup.
   - Return to your seat. **Exchange #1** has been completed.

3. WAIT until all the students in the class have completed Exchange #1.

   WHEN YOUR INSTRUCTOR TELLS YOU TO START, perform **Exchange #2** using the same procedures as you did for Exchange #1.

   Select a partner from another area in the classroom. DON'T SELECT THE SAME PARTNER YOU HAD BEFORE.

4. Pour a **small amount** of liquid from your cup of "body fluids" into a clean sampling cup (just enough to **cover the bottom** of your sampling cup). Add **one drop** of **phenolphthalein** indicator.

   If the liquid in your sampling cup **turns bright pink**, you've been **infected with HIV.**

5. Based on the results of your indicator test, check the appropriate box in **Table 1-4** (infected or not infected) for Exchange #2.

   After recording the results of your indicator test, **dispose** of the **liquid and the sampling cup.**

TABLE 1-4 RESULTS OF INDICATOR TESTS FOR BODY FLUID EXCHANGES		
	INFECTED	NOT INFECTED
Exchange #2		✓
Exchange #4		✓
Exchange #6		
Distilled water		✓

6. Using the dropper bottle of distilled water, drop a **small amount of distilled water** into a clean sampling cup (just enough to **cover the bottom** of your sampling cup). Add **one drop** of **phenolphthalein** indicator.

   Based on the results of your indicator test, check the appropriate box in **Table 1-4** (infected or not infected) for distilled water.

7. What was the reason for performing the phenolphthalein indicator test on the **distilled water? Explain** your answer.

   as a control, or negative control

8. In scientific terminology, what name is given to the part of our experiment in which we performed an indicator test on the distilled water? _____

9. WAIT until all the students in the class have completed the indicator tests for distilled water.

    WHEN YOUR INSTRUCTOR TELLS YOU TO START, perform **Exchange #3.**

    Select a partner from another area in the classroom. DON'T SELECT ANY OF THE PARTNERS YOU HAD BEFORE.

10. WAIT until all the students in the class have completed Exchange #3.

    WHEN YOUR INSTRUCTOR TELLS YOU TO START, perform **Exchange #4.**

    Select a partner from another area in the classroom. DON'T SELECT ANY OF THE PARTNERS YOU HAD BEFORE.

11. Perform the indicator test on the "body fluids" in your paper cup, using a clean sampling cup.

    Based on the results of your indicator test, check the appropriate box in **Table 1-4,** but this time you're recording the results for Exchange #4.

    After recording the results of your indicator test, **dispose** of the **liquid and the sampling cup.**

12. WAIT until all the students in the class have completed the Exchange #4 indicator tests and recorded their results.

    WHEN YOUR INSTRUCTOR TELLS YOU TO START, perform **Exchange #5.**

    Select a partner from another area in the classroom. DON'T SELECT ANY OF THE PARTNERS YOU HAD BEFORE.

13. WAIT until all the students in the class have completed Exchange #5.

    WHEN YOUR INSTRUCTOR TELLS YOU TO START, perform **Exchange #6.**

    Select a partner from another area in the classroom. DON'T SELECT ANY OF THE PARTNERS YOU HAD BEFORE.

14. Perform the indicator test on the "body fluids" in your paper cup, using a clean sampling cup.

   Based on the results of your indicator test, check the appropriate box in **Table 1-4**, but this time you're recording the results for **Exchange #6.**

   After recording the results of your indicator test, **dispose** of the **liquid and the sampling cup.**

15. When everyone has recorded his/her results for Exchange #6, your instructor will survey the class to determine the total number of infected students after **Exchanges #2, #4, and #6.** Record the results of the survey in **Table 1-5.**

TABLE 1-5 SIMULATION RESULTS — SPREAD OF HIV	
EXPERIMENTAL TRIALS	NUMBER OF INFECTED CLASS MEMBERS
Start of experiment	1
Exchange #2	
Exchange #4	
Exchange #6	

ACTIVITY 5   GRAPHING YOUR RESULTS

1. Graphs provide a good visual representation of the relationships between the factors investigated in an experiment.

2. Look at the graph structure in **Figure 1-1** and note the following **key points:**

    ■ The **horizontal** axis is referred to as the *X*-**axis.** The **vertical** axis is called the *Y*-**axis.**

    ■ Numbers on the X- and Y-axes must have an **equal interval** between them (for example, 5, 10, 15 but **not** 5, 10, 20, 50).

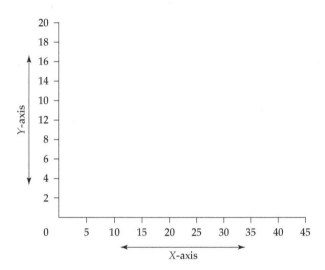

FIGURE 1-1. Graph Structure

■ When deciding which factors to plot on the X- and Y-axes, here's a good rule of thumb: if one data set consists of words (months of the year, car models, country names) and the other is numbers (number of cars sold, average income), plot the **words on the X-axis** and the **numbers on the Y-axis.** If both sets of data are numbers, plot the **factor being measured** on the Y-axis.

■ Numbers on the X- and Y-axes are chosen carefully to make **best use** of the space available. It's not necessary to always begin numbering with zero. It's also not necessary to use the same number scale on both the X- and Y-axes. As shown in **Figure 1-1,** each axis should have a number scale tailored specifically to the data being presented.

■ It's not permissible to extend lines or bars **outside the margins** of the graph. Adjust the graph scale to make the data fit comfortably.

3. **Bar graphs** and **line graphs** are examples of types of graphs that are used frequently to present scientific data. **Figure 1-2** illustrates two different ways to present the same information. Note that in both versions of the graphed data:

    ▪ Lines or bars are **large and easy to read.**

    ▪ Each graph has a **title** that describes the subject matter being graphed. The title can be placed either above or below the graph.

    ▪ Each axis has a **title** that clearly explains the numbers listed there.

    ▪ If the graph contains more than one set of bars or lines (as is the case in **Figure 1-2**), each must be identified with a **key.**

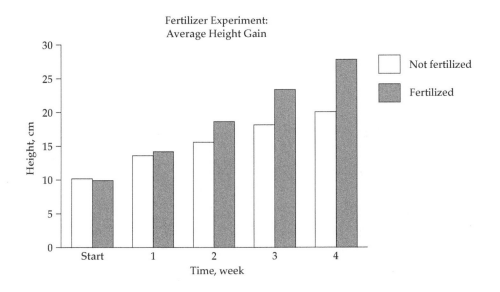

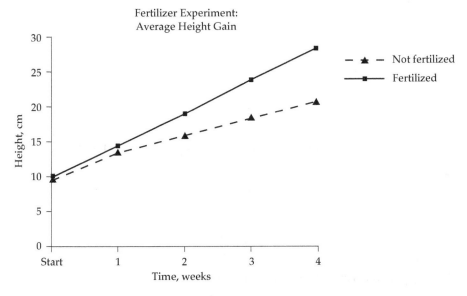

FIGURE 1-2. Comparison of Bar and Line Graphs

✔ Comprehension Check

1. On the graph paper in **Figure 1-3, plot a graph of your experimental results.**

   Choose an appropriate number scale for the **Y-axis** and label it **Number of Infections.** Choose an appropriate number scale for the **X-axis** and label it **Number of Exchanges.**

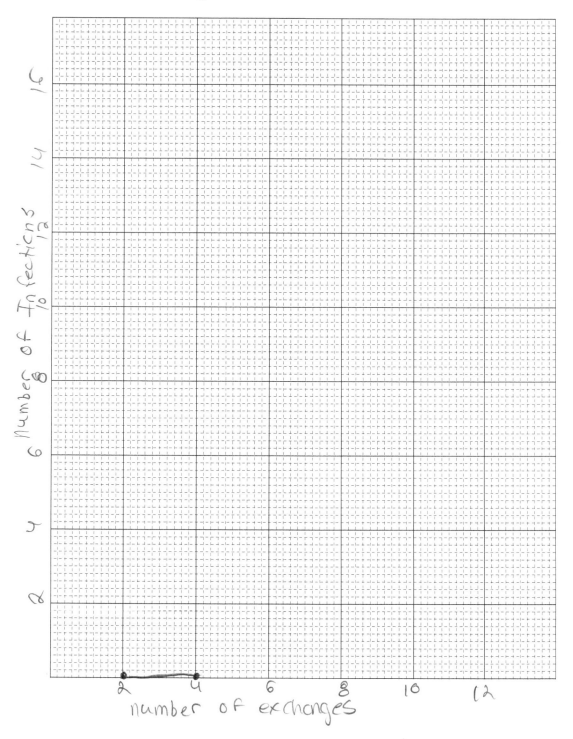

FIGURE 1-3. Results of HIV Simulation

2. In a few sentences, summarize the results of the experiment. In your summary, include data from your summary chart and graph. **Statements of results should include only facts — no interpretation.**

In exchange #2 #4 we did not get In fected.

3. Based on your graph, estimate the number of body fluid exchanges it would take for the whole class to become infected. _____O_____ exchanges

4. In real life, infections don't spread as rapidly as they did in our simulation? Why not?

Because in a class room setting we are in closed more, and confined to each other. In the real world you can spread out more and became less likely to became Infected.

5. Suggest some methods to slow the rate of infection in the general population.

use proper PPE
If you are Infected let people know.

Check your graph and answers with your instructor before you continue.

ACTIVITY 6 INTERPRETING DATA USING SIMPLE STATISTICS

Often, the data collected in an experiment is in a form that isn't easily understandable. Measurements have been established to make it easier to interpret and draw conclusions from large collections of information. We're all familiar with the U.S. Census, which collects huge amounts of information on family size, income, housing conditions, population distributions, and so on.

Simple statistical analysis can reduce the data and convert it to a usable form. A similar approach is used when analyzing the results of large experiments (such as evaluating the effectiveness of new medications or airbags in automobiles). The most commonly used statistical measures are the mean, the mode, the median, the range, and the standard deviation.

The **mean** is the **average** of a set of numbers. The mean is equal to the **sum of all the numbers in the set divided by the sample size.** For example, to find the average pulse rate of a group of ten students, you would add the pulse values for all of the students together and then divide the answer by the number of students (10).

The mean of a group of numbers often doesn't really give you the information you need to correctly interpret the data. **For example, the following two sets of numbers have exactly the same mean, but the spread (dispersion) of numbers is quite different.**

Set 1: 39, 38, 38, 40, 40 mean = 39

Set 2: 3, 29, 25, 38, 100 mean = 39

The **mode** is the **most frequently occurring number** in a set. The mode represents the most common response and, therefore, can be used as a predictor to determine market response (for example, which car model will sell the best in a specific area of the United States).

The **median** is the **middle** number of a set when they're arranged in either **ascending or descending order.** If your income level is above the median, for example, your salary is in the upper 50% of salaries being compared. If a set of numbers has **no middle value,** you can find the median by **averaging the middle two numbers** in the set.

The **range** is the **difference between the largest and smallest values** in a set, for example, the difference between the number of yards gained by the best and worst running backs in the National Football League.

To demonstrate how applying different statistical measures changes the meaning of results, consider the set of 20 biology exam scores in **Table 1-6**.

TABLE 1-6			
BIOLOGY EXAM SCORES			
STUDENT NUMBER	SCORE: EXAM 1	STUDENT NUMBER	SCORE: EXAM 1
1	90	11	88
2	94	12	54
3	80	13	32
4	82	14	47
5	91	15	25
6	46	16	56
7	97	17	59
8	96	18	60
9	87	19	87
10	84	20	86
Mean = 72.05			
Mode = 87			
Median = 83			

1. What's the **range** of exam scores in **Table 1-6?**

    72

2. If your **exam score was 80,** was your score in the top 50% of the class? **Explain** your answer.

    NO, because you are lower than the median of 83

3. In this situation, is the **mean** a good representation of the class scores? **Why or why not?**

    Yes, because its the average. The outlyer are not effecting the results.

4.   While watching television last night, you saw an advertisement for Lose-Fast Weight Control Pills. The 12 women shown lost an **average** of 30 pounds while taking the pills. What additional **statistical measures** (as discussed in **Activity 6**) should be reported for you to make an informed decision whether or not to purchase this product?

length of experiment

what weight did they start at

medical conditions they have

    Check your answers with your instructor before you continue.

Self Test

For each sentence below, enter the letter of the correct step of the scientific method.

a.   **Test hypotheses (by experiment or observation)**
b.   **State hypotheses**
c.   **State results (facts only)**

1.   _B_   seeds will grow faster if you fertilize the ground before you plant them.

2.   _C_   in an experiment, 70 of 80 household cockroaches were attracted to peanut butter.

3.   _b_   Tanya grew bacteria from her mouth on special plates in the laboratory. She placed drops of different mouthwashes on each plate.

4.   _a_   Kevin designed a survey to determine how many of his classmates had dimples on their chins and how many did not.

5.   _a_   plants grown under red light will grow faster than those under white light.

6.   _b_   if acid rain affects plants in a particular lake, it might also affect small animals that live in the same water.

7.   _B_   Maria's experiment showed that chicken eggshells are more resistant to crushing when the hens are fed extra calcium.

Identify the graphing mistakes in **Figures 1-4, 1-5, and 1-6:**

8.

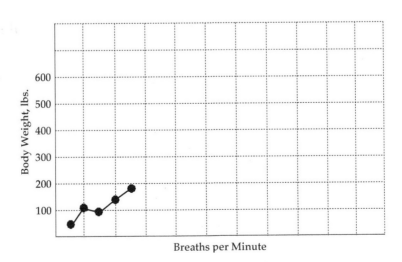

FIGURE 1-4.   Sample Graph One

9.

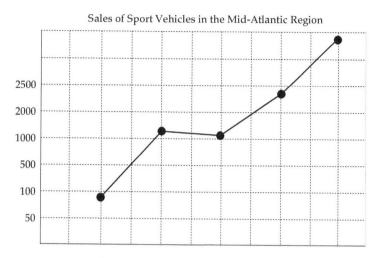

FIGURE 1-5. Sample Graph Two

10.

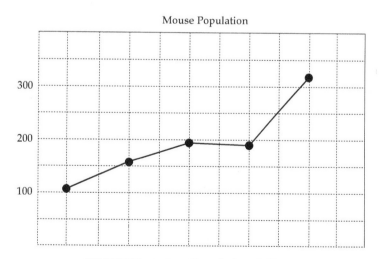

FIGURE 1-6. Sample Graph Three

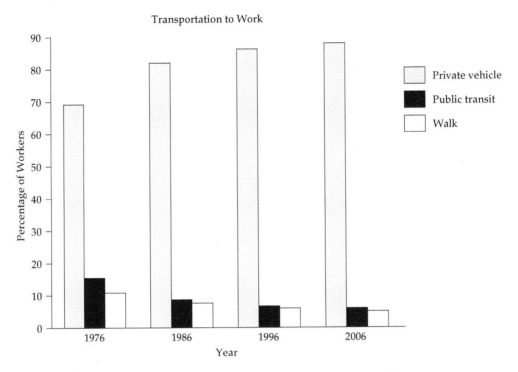

FIGURE 1-7. Percentage of Workers Using Various Modes of Transportation

Answer the following questions in reference to the graph in **Figure 1-7.**

11. **(Circle one answer.)** Between 1976 and 2006, the percent of workers who walked to work **increased / decreased / remained the same.**

12. Between 1976 and 1986, the number of people who drove cars to work increased by _____ %.

13. In 2006, what percent of workers drove cars to work? _____.

14. The number of workers who used public transportation dropped by 7% between 1976 and _____.

Due next week

Using the Compound and Dissecting Microscopes

Objectives

After completing this exercise, you should be able to:

- identify the parts of the compound and dissecting microscopes and explain their functions
- choose the correct type of microscope for viewing different specimens
- focus the compound microscope using the scanning, low-, and high-power lenses
- prepare a wet mount slide
- correct viewing problems that commonly occur when using the compound microscope
- accurately describe specimens viewed through the dissecting and compound microscopes
- use the microscope to test a hypothesis

CONTENT FOCUS

All organisms, large and small, make valuable contributions to the functioning of the ecosystem. Many plants and animals are large and easy to see, but many important organisms are too small to be seen without the assistance of magnifying lenses.

You'll have an opportunity to get a close look at the anatomy and behavior of some interesting organisms that live around you unnoticed. Microscopes work like magnifying glasses to give you a closer look at these small organisms.

Different types of microscopes can be used for different purposes. During this exercise, you'll be learning to use a **dissecting microscope** to examine larger objects and a **compound microscope** to view smaller specimens.

23

ACTIVITY 1

LEARNING TO USE
THE DISSECTINGMICROSCOPE

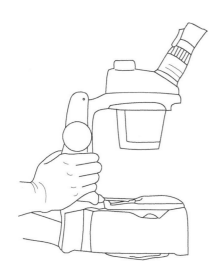

> ### Caution!
> Dissecting microscopes may have several parts that are not permanently attached!

1. Work in groups of **two students.** Get the following supplies: **a dissecting microscope and a penny.**

2. Place the penny on the microscope **stage.** Locate the stage by referring to the diagram in **Figure 2-1.**

    Turn the penny so that the Lincoln Memorial is facing you. **Adjust the light** until the penny is brightly illuminated.

3. Set the **magnification control knob,** located on the top or the side of the **head,** to the **lowest setting.**

4. Turn the **focusing knob,** located on the microscope **arm,** until the head is as close to the stage as possible.

    Look through the **ocular lenses (eyepiece)** at the penny. **Turn the focusing knob** until the image of the penny is sharp and clear (the head will be moving **away** from the stage).

    Is Lincoln sitting in the Lincoln Memorial in your penny? *No*

5. While looking through the eyepiece, gradually turn the magnification control knob. The change in image size will resemble the zoom action of a camera.

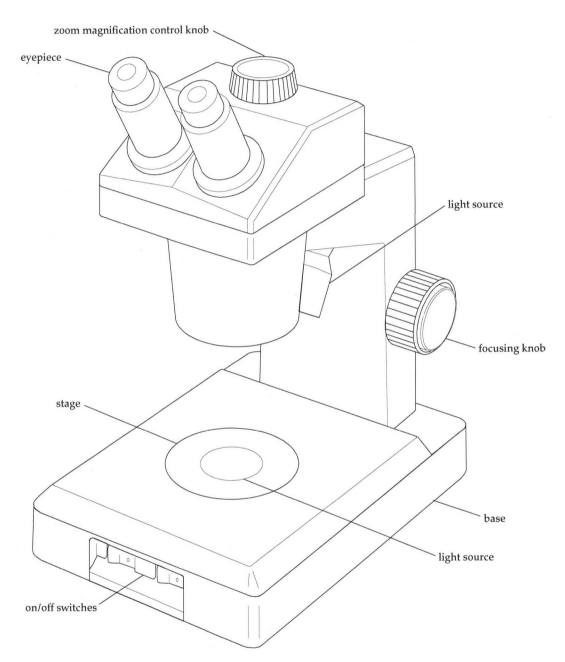

FIGURE 2-1. Parts of the Dissecting Microscope

ACTIVITY 2

MAKING OBSERVATIONS WITH THE DISSECTING MICROSCOPE

1. Work in groups. Get the following supplies: **a small glass bowl, a pipette, and a blunt metal probe.**

2. Use the pipette to remove a **planaria** from the culture container and place it in your bowl. Locate the worm of your choice and draw it into the tip of the pipette (as shown in **Figure 2-2**). When the planaria feels the water current, it will probably form itself into a small protective ball, and it will be easy to draw it into the pipette.

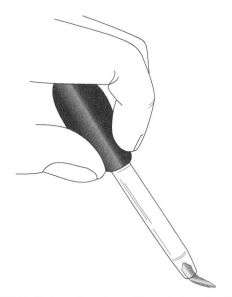

> ## Caution!
>
> **Don't suck the planaria too far into the pipette. It will attach to the inside of the pipette and you won't be able to get it out.**
>
> **If additional water is needed in your glass dish, add pond water. Don't use tap water or distilled water.**

FIGURE 2-2. Drawing a Planaria into the Pipette

3. **Observe your worm** under the **dissecting microscope.** Use a medium or low light level.

    Observe carefully through the microscope for at least **two minutes.**

    Does the worm swim or crawl? _Swim_

4. **Draw a picture** of your worm. **Increase the magnification** in order to see the details of the head. When drawing the head, pay particular attention to the position of the eyes in relation to each other. Make your drawing **large** and **clear. Label** the **head** and the **eyes.**

DRAWING OF PLANARIA

5. Planaria are common inhabitants of freshwater ponds and streams. Based on your **observations**, how do you think they get their food? **Explain** your answer.

   Obsorbing through the water

6. Hold the **blunt probe motionless** directly in the path of a moving worm. How does the worm respond?

7. Touch the head end **gently** with the blunt probe. How does the worm respond?

8. Touch the planaria **gently** on several other body parts. Record your observations.

9. In what way was the **touch response** on other body parts **similar or different** to the response you observed when the worm was touched on the head end?

Check your answers with your instructor before you continue.

ACTIVITY 3   GETTING FAMILIAR WITH THE COMPOUND MICROSCOPE

1. Work in groups. Get a **compound microscope and a blunt probe.**

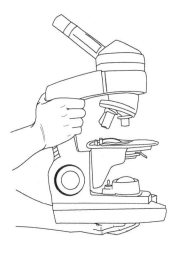

> ### Caution!
> **Always carry the microscope upright with one hand under the base and the other hand on the arm.**

2. Plug the microscope into an electrical outlet and **turn the light on.** The light switch is located on the **base** of the microscope.

3. The compound microscope consists of a system of **optics** (lenses and mirrors) and focusing controls. The base and the arm support a **body tube** that **houses the lenses** that magnify the image.

4. At the top of the scope is the **ocular lens (eyepiece).** The ocular lens is only one of a series of lenses that magnify the image. The ocular lens makes the image **ten times larger** than life size (abbreviated **10×**).

   Look through the ocular lens. You'll see a black pointer. Rotate the eyepiece **while looking through the lens.**

   What happens to the pointer? ___*Stays still*___

5. At the bottom of the body tube is a revolving **nosepiece.** The lenses that screw into the nosepiece are called **objective lenses.**

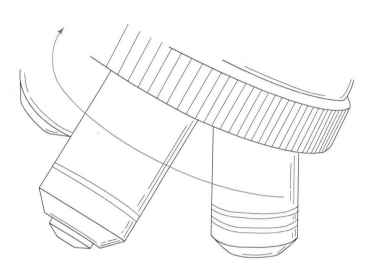

Turn the nosepiece until you hear or feel one of the objective lenses quietly **click into position** (see **Figure 2-3**). When an objective lens clicks into position, it's in the proper alignment for light to pass from the light source, through the objective lens, through the ocular lens, and into the viewer's eye. Turn the nosepiece again to bring a different objective lens into position.

FIGURE 2-3.   Revolving Nosepiece and Objective Lenses

6. **How will you know** when you've placed the objective lens in the proper position?

   Locks into place and you can see the object

7. Note that each objective lens is a different length. Each of the lenses has a different magnifying power. The **shortest lens** has the **lowest magnifying power** and the **longest lens** has the **highest magnifying power.**

   Since light from the specimen passes through **both** an objective lens **and** the ocular lens, the **total magnification** of the image is the result of the objective lens magnification **multiplied** by the ocular lens magnification.

8. Using the information provided in step 4, **complete the magnification table.**

Length of Objective Lens	Magnification of Objective Lens	Magnification of Ocular Lens (Eyepiece)	Total Magnification of Specimen
Short	4×		
Medium	10×		
Long	40×		

9. On both sides of the arm, you'll see the **focus knobs.** The larger ring is the **coarse focus knob** and the smaller ring is the **fine focus knob.**

   Turn the microscope so that you're looking at it **from the side.**

   Position the **blunt probe vertically** at the side of the microscope so that the **tip** of the probe is approximately **level with the stage** of the microscope.

   Turn the **coarse focus knob** all the way in one direction and then reverse the process. **What happens to the position of the stage, relative to the tip of the probe?**

10. Repeat the process with the probe, but this time, **rapidly** turn the **fine focus knob** all the way in one direction and then reverse the process. **What happens to the position of the stage, relative to the tip of the probe?**

11. The **mechanical stage** is a movable platform designed to hold a microscope slide. Notice that one side of the metal stage clip makes a 90° angle. When you place a slide on the stage, **make sure** that one corner of the slide **fits exactly into that angle.** If not, you won't be able to move the slide or focus properly.

    Two control knobs move the mechanical stage. One moves the stage left and right; the other moves the stage forward and back. Always move the stage using the control knobs. Don't attempt to move the slide using your fingers.

12. In the center of the stage, you'll see the glass of the **condenser lens,** which focuses the light on the specimen.

13. Directly above the light source, you'll see a lever that moves from left to right. This lever is connected to the **iris diaphragm.** Looking at the condenser lens, move the **iris diaphragm lever** all the way to the **left. What happens?**

Moves away from you

14. Looking at the side of the microscope, move the **iris diaphragm lever** all the way to the **right. What happens now?**

move closer to you

ACTIVITY 4   PARTS OF THE COMPOUND MICROSCOPE

Now that you have some experience with the parts of the microscope and their functions, **label the parts** of the microscope on **Figure 2-4**.

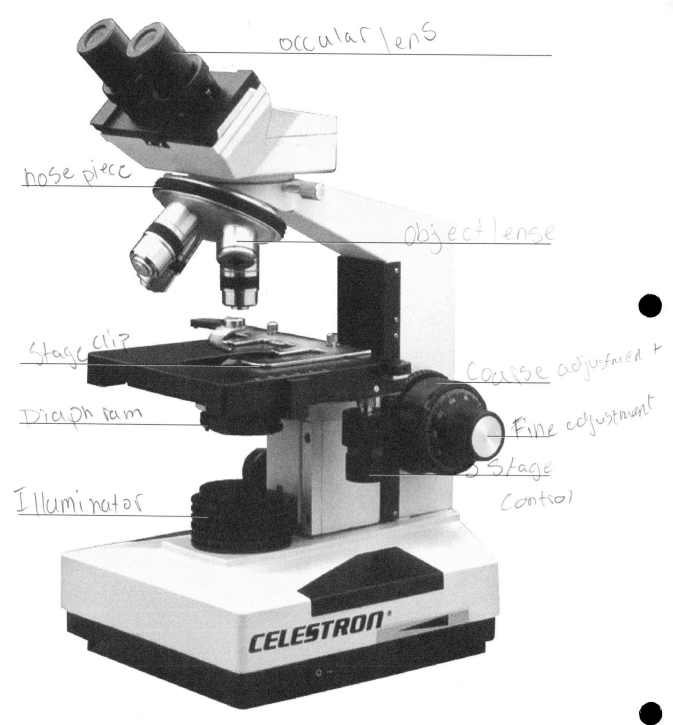

occular lens

nose piece

object lense

Stage clip

Course adjustment

Diaphram

Fine adjustment

Stage control

Illuminator

FIGURE 2-4.   Parts of the Compound Microscope

LEARNING TO USE
ACTIVITY 5
THE COMPOUND MICROSCOPE

1. Work in groups. Get the following supplies: **a prepared slide labeled letter "e" and a piece of lens paper.**

2. Complete the following steps to view the letter "e" (or another specimen) with the compound microscope:

    ▪ Plug in the microscope and turn on the light. Open the iris diaphragm so that you can see the light shining through the condenser lens.
    ▪ Using the coarse focus knob, lower the stage as low as it can go.
    ▪ Clean the slide with lens paper and carefully position the slide so that it fits precisely into the 90° angle of the metal clip on the stage. Position the slide on the stage so that the letter "e" is facing you **right-side up,** in its **normal reading position.**
    ▪ Rotate the **lowest power** objective lens into position.
    ▪ Using the control knobs **on the mechanical stage,** position the slide so that light is shining on the letter "e."
    ▪ While looking through the eyepiece, use the **coarse focus knob** to raise the stage slowly. Continue until the letter "e" **pops into focus.**
    ▪ Once again, use the **control knobs** on the mechanical stage to **center** the letter "e" in your field of view.
    ▪ Using the **fine focus knob,** adjust the focus until the letter "e" is sharp and clear.

✓ Comprehension Check

1. If the letter "e" isn't lighted brightly enough for you, **what should you do?**

2. If the light is too bright, causing glare and eyestrain, **what should you do?**

3. What is the **total magnification** of the image of the letter "e"? **Show your work.**

Check your answers with your instructor before you continue.

3. Draw the letter "e" **exactly** as it appears under low power. Make it **large** and **clear**.

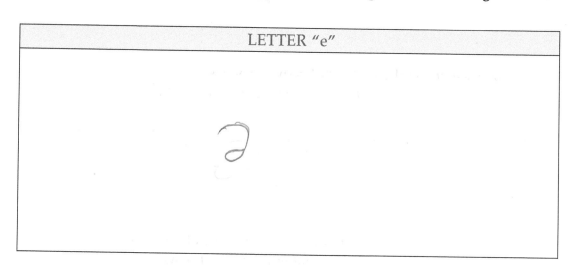

LETTER "e"

4. How is the **orientation** of the letter "e" as seen through the microscope **different** from the way an "e" **normally** appears? List **two** differences.

    It is reversed.

    Ink looks blotted

5. **While looking through the eyepiece,** move the stage to the **left.**

    In what direction does the image appear to move? *move to the right*

    While looking through the eyepiece, move the stage **away from you.**

    In what direction does the image appear to move? *to the top*

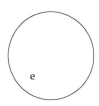

e

If you wanted to **center** the letter "e" in this drawing, in which **two directions** would you have to move the stage?

    right   and *towards you*

6. Follow these steps to view a specimen at a higher magnification:

   - Center the letter "e" (or other specimen) in your field of view.
   - Rotate the **10×** objective lens into position. **DON'T adjust the stage at this point!**
   - **DON'T use the coarse focus knob!** Using the fine focus knob, adjust the focus **slowly** until the letter "e" is sharp and clear.
   - Repeat the above steps to change to other objective lenses.

7. **Centering the specimen is** absolutely necessary **before you can change to a higher-power objective!** Do you know why? The reason is simple. The more powerful the magnifying lens, the smaller the area you see, but you see that small area in greater detail. The area you can see at one time is called the **field of view.** So you could say that the **higher** the power of the objective, the **smaller** the field of view.

   If the specimen you're trying to view isn't centered before you switch to a higher-power lens, it may no longer be within the field of view, and you'll think your specimen has disappeared!

8. Looking through the eyepiece, can you still see the letter "e"? __Yes__

   Is the letter "e" exactly in the center of the field of view? __Yes__

   If not, **move the slide slightly** to center the image.

   Is the letter "e" sharp and clear? __Yes__

   If not, **gradually adjust the fine focus knob** until the problem is corrected.

   Do you have enough light? __Yes__

   If not, **gradually adjust the iris diaphragm.**

   How has the **image of the letter "e" changed** from the way it looked using the 4× objective lens? __You can see all the ink particals. Little dots all connecting__

9. Repeat the instructions in **step 6** above, this time changing from **10×** to the **high-power objective.**

   Looking through the eyepiece, can you see the entire letter "e"? __Yes__

10. **(Circle one answer.)** Your field of view on high power is **larger** / **smaller** than the field of view on 10×.

   How does the size of the field of view determine how much of the letter "e" you can see? *The lower maginficatirey the more you see of the letter*

11. When using the **high-power objective,** 40× what is the **total magnification** of the image of the letter "e"? *400* _____ **Show your work.**

   40×10 = 400

  Check your answers with your instructor before you continue.

ACTIVITY 6   PREPARING TEMPORARY SLIDES — WET MOUNTS

A wet mount is a method of preparing a slide that will be used only for a short time. Unlike the letter "e," which was permanently attached to the slide, a wet mount is made by placing the specimen into a drop of liquid on a slide. The specimen and water droplet are held in place by a coverslip.

Human Epithelial Cells (Cells from the Inside of Your Cheek)

1. Work in groups. Your group will make **two different wet mounts** of cheek cells.

   One will be made with a drop of **stain.** The second will be made by substituting a drop of **physiological saline** for the stain.

2. Get the following supplies: **a dropper bottle of iodine stain or physiological saline, a slide, a coverslip, lens paper, and a clean toothpick.**

3. **Place a drop of iodine stain or a drop of saline near the center of a clean slide.**

4.   **Gently** scrape the inside of your cheek with the **end of a toothpick.**

Caution!
If you scrape too hard, you'll be examining blood cells instead of epithelial cells!

You've removed some of the cells that form a **protective covering** for the inside of the mouth. Like other epithelial cells, these are constantly being worn off and replaced by new cells of the same type.

5.   **Spread the material from the toothpick** into the drop of iodine stain or saline. Add a coverslip.

     Put two microscopes together (yours and your partner's) on the laboratory table so that you can view and compare the stained and unstained slides.

Hint:
Unstained cells are clear. They're only visible with very low light levels. If you think there are no cells on your slide, adjust the iris diaphragm.

6.   **View both slides. Begin with the 4× objective.** Continue until you've located the cells using **all three objective lenses.**

7.   **Draw a picture** of **one** cheek cell, **viewed on high power.** Make it **large and clear.**

HUMAN CHEEK CELL

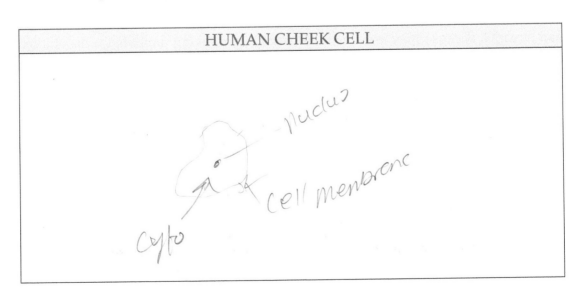

8. **Label** the following cell structures (referred to as **organelles**) in your drawing of the cheek cells:

Organelles to Label	Function
nucleus	directs all cell activities
cell membrane	controls movement of materials in and out of the cell
cytoplasm	jelly-like fluid found between the nucleus and cell membrane

9. Give an example of a **medical procedure** in which epithelial cells are scraped from another area of the body and **examined microscopically.**

    Swabbing for strep throat

10. Is there an advantage in **using a stain** to view cells microscopically? *No* If so, how is the stain helpful?

    to much stain makes it so you cant see anything

Daphnia — An Aquatic Organism

The water flea, daphnia, is a microscopic organism commonly found in ponds, lakes, and streams. Because they are small and transparent, living daphnia can be studied easily in the laboratory. Daphnia feed on microscopic food particles. Their five pairs of legs are modified into strainers that filter the food particles from the water. Daphnia, in turn, are an important part of the food chain. Many fishes and even larger aquatic animals feed on daphnia.

1. Work in groups. Get the following supplies: **a depression slide and a bottle of methyl cellulose** (the bottle may also be labeled **Protoslo®**).

2. Put a **small** drop of the methyl cellulose into the depression on your slide.

    Use the pipette in the culture jar to remove a daphnia from the container and place it in the depression on the slide. **You don't need a coverslip.**

3. Make your observations using the **scanning lens** of the microscope (**4×**). Add the following details to the daphnia outline below: **eye, heart, and intestine**.

   Label the details you're adding to the drawing. Also, label the head and legs.

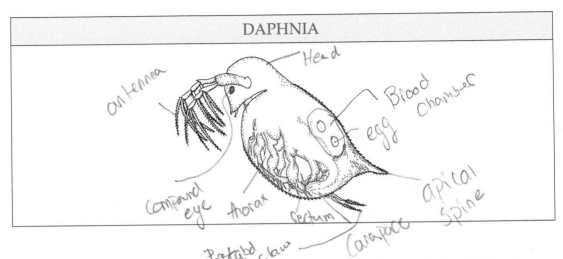

DAPHNIA

4. To see details of their internal structure, both the daphnia and the cheek cell must be viewed through the microscope. Since they're both so small, why were you able to see organs such as the brain and the heart in the daphnia, but not in the cheek cell? **Explain** your answer.

   The cheek cells don't have a brain or heart they are just basic cells.

Self Test

1.   Complete the table by entering the appropriate part of the compound microscope
      or the correct function for the part of the microscope listed.

PARTS OF THE COMPOUND MICROSCOPE	FUNCTION
Ocular lens (eyepiece)	
Stage	
	Focuses light on the specimen
Nosepiece	
	Objective lens used to first locate a specimen
	Regulates the amount of light that passes through the specimen
Fine focus knob	
	Objective lens with the lowest magnifying power
	Objective lens with the highest magnifying power
Coarse focus knob	

2.   What is the **total magnification** if the ocular lens is **15×** and the objective lens is
      20×? Show your work.

3. List **three differences** between the dissecting microscope and the compound microscope:

    a.

    b.

    c.

4. **What type of microscope** (compound or dissecting) would you use to observe the following:

    _____ Cells from the lining of your stomach

    _____ A splinter in your finger

    _____ A cockroach you found in the kitchen

    _____ Mold from your shower curtain

5. Suppose you were watching a daphnia under the compound microscope and noticed that it moved **toward you** and then **to your right.**

    Which direction(s) did the daphnia **actually** move? _____

    Explain your answer.

6. Explain how you would **correct the following problems** experienced when using a microscope:

   a. When changing magnification from **10×** to **40×**, the specimen disappears.

   b. The field of view is too dark.

   c. Your field of view is partially obscured by a dark area.

   d. There's a fingerprint in your field of view.

   e. There are many hollow, dark circles in your field of view.

Functions and Properties of Cells

Objectives

After completing this exercise, you should be able to:

- identify and explain the functions of the major cellular organelles
- explain the similarities and differences between plant and animal cells
- explain the concepts of diffusion and osmosis and why they are important to cell physiology
- use indicator chemicals to test for the movement of molecules and to determine the direction of diffusion
- explain the process of osmosis in living cells exposed to different extracellular solute concentrations
- apply your knowledge of cell structure and function to real-life situations

CONTENT FOCUS

Typically, when you look at a plant or animal, it's easy to think of an organism as one large unit. When looking through the microscope, however, it becomes obvious that an organism is composed of trillions of tiny units called **cells**, which work cooperatively to carry out the functions that keep us alive. If the activity of your cells stopped, even for a few moments, death could quickly follow.

There are about 200 different types of cells in the human body. These cells can be quite different in shape, structure, and function, but they all have some basic characteristics in common.

Within each cell lies a collection of specialized structures called **organelles,** in which particular chemical activities take place. The various cell structures carry out activities that mirror the functions of our body organs. Most organelles are compartments bounded by membranes.

There several ways that nutrients, ions, oxygen, carbon dioxide and other molecules can be moved in and out of cells across the cell membrane. Two of these transport methods, **diffusion** and **osmosis**, take advantage of the fact that molecules are in constant motion.

The process of **diffusion** occurs whenever dissolved particles move from an area of **high concentration** (more of them) to nearby areas where they are **less concentrated.**

Cells are surrounded by a membrane that allows some substances to pass through, but not others. This is referred to as a **selectively permeable membrane.**

A process similar to diffusion occurs with **water** molecules in the environment. Water molecules disperse through a **selectively permeable membrane** from an area of **high concentration** to an area of **lower concentration.** This process is given a different name, **osmosis.**

ACTIVITY 1   CELLULAR ORGANELLES

Learning about cell organelles will be more meaningful when you understand the role that each part plays in the life of a cell. Often the presence of organelles is **related to the function** of various cells. In a cell with a specific function, **some organelles may be more numerous** than in other nearby cells with different functions.

Some cell organelles are large enough to be seen with your compound microscope. Others are much smaller, but can be seen and photographed with more powerful microscopes. Also, although **most organelles are found in both plant and animal cells**, there are some **exceptions**.

Table 3-1 gives a brief summary of the functions of some basic organelles. Based on the information in the table, **fill in the blanks with the appropriate organelle or organelles.** Answers can be used **more than once.**

1. Three organelles that are found in plant cells, but not in animal cells: _Chloroplast_ , _Cell wall_ , and _Central vacuole_

2. Organelle where muscle proteins are manufactured. _Rough ER_

3. Provides strength and support in tree trunks. _Cell wall_

4. An infertile man has a sperm sample tested and discovers the sperm have low motility (don't swim normally). Name the malfunctioning organelle. _Cillia or flagela_

5. Two structures that a material would have to cross to enter the cytoplasm of a plant cell. _cell wall_ and _cell membrane_

6. Would be more numerous in muscle cells than in skin cells. _mitochondra_

7. Would be functioning in a fat storage cell. *Smooth ER*

8. Location where genetic information is stored. *nucleus*

9. Location in the cell where most organelles can be found. *Cytoplasm*

10. Allows some materials to enter the cell, but not others. *Cell membrane*

T A B L E   3 - 1
SOME CELL ORGANELLES AND THEIR FUNCTIONS

ORGANELLES VISIBLE WITH A LIGHT MICROSCOPE	FUNCTION
Cell wall	External support and protection; if present, located outside the cell membrane; composed primarily of cellulose; not found in animal cells
Cell membrane	Surrounds the cytoplasm; barrier between the environment and the cytoplasm; controls the movement of materials in and out of the cell
Cytoplasm	Liquid/gel "filler" substance inside the cell membrane; all internal cell organelles are suspended in the cytoplasm
Nucleus	Stores and transfers information (DNA and RNA) needed to control cell functions
Chloroplast	Structure in which photosynthesis takes place; contains chlorophyll and is green in color
Central vacuole	Membrane-enclosed bag of fluid; water storage organelle in a mature plant cell; not found in animal cells
Cilia or flagella	Hair-like projections from the cell membrane; capable of movement; assists in cell locomotion or moves materials across the cell surface
ORGANELLES NOT VISIBLE WITH A LIGHT MICROSCOPE	FUNCTION
Mitochondria	Location of aerobic cellular respiration; produces ATP energy, which is used by cells to do work
Ribosomes	Protein synthesis; ribosomes may be found free in the cytoplasm or attached to the endoplasmic reticulum (ER)
Endoplasmic reticulum (ER)	Network of membranes important for transport of molecules within the cytoplasm and across the cell membrane; many chemical reactions occur here, including synthesis of some carbohydrates, lipids, and proteins; divided into "rough" and "smooth" sections with different functions
Smooth ER	Carbohydrate and lipid synthesis
Rough ER	Protein synthesis; ribosomes are attached to the ER membrane in this location
Golgi apparatus	Packaging of molecules for export out of the cell; most exported molecules are proteins

✓ Comprehension Check

1. In an accident in a chemical plant, cyanide gas was accidentally released. It's known that exposure to cyanide affects production of the energy-rich molecule ATP. Which cellular **organelle** would be affected by the poison? **Explain** your answer.

Check your answers with your instructor before you continue.

ACTIVITY 2   OBSERVING LIVING CELLS

1. Work in groups. Get **two compound microscopes** and set them up side by side.

2. **Amoeba** is a **one-celled organism** commonly found in pond water.

    Elodea is a **freshwater plant** frequently used in home aquariums. You'll be looking at an elodea leaf, which is thin enough to use with the compound microscope. An elodea leaf has two layers of cells. Each cell looks like a small rectangle.

    Following directions from your instructor, make **two wet mounts,** one of an **amoeba** and one of an **elodea leaf.** Place **one wet mount on each microscope.**

3. **Figure 3-1** contains a photograph of an amoeba. Based on your microscope observations and the photo in the figure, **label** all the organelles you can identify on the **photo of the amoeba.**

4. **Figure 3-2** contains photographs and a diagram of elodea cells. Based on your microscope observations and the photos in the figures, **label** all the organelles you can identify on **both the diagram and photo of the elodea.**

Hint:

Use the information in Table 3-1 to help you locate and identify the cellular organelles.

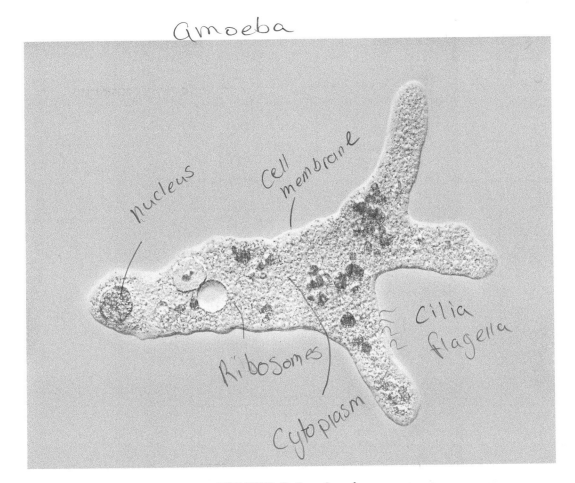

FIGURE 3-1.   Amoeba

FIGURE 3-2A. Elodea Leaf

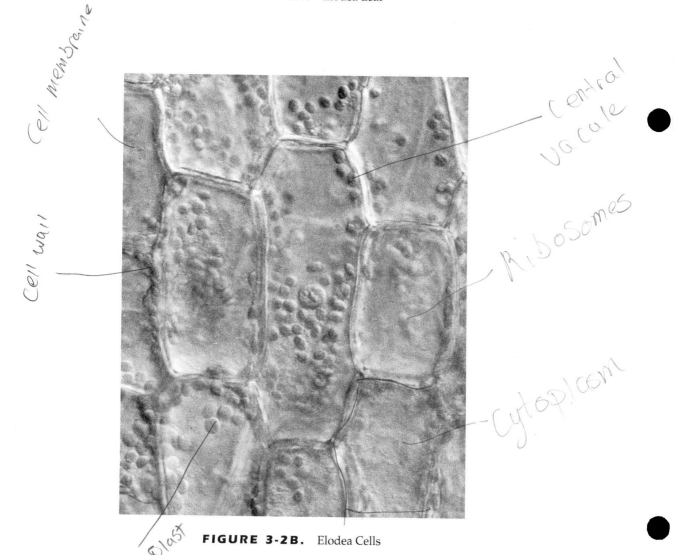

Cell membrane

Cell wall

Choloplast

Central Vacule

Ribosomes

Cytoplasm

FIGURE 3-2B. Elodea Cells

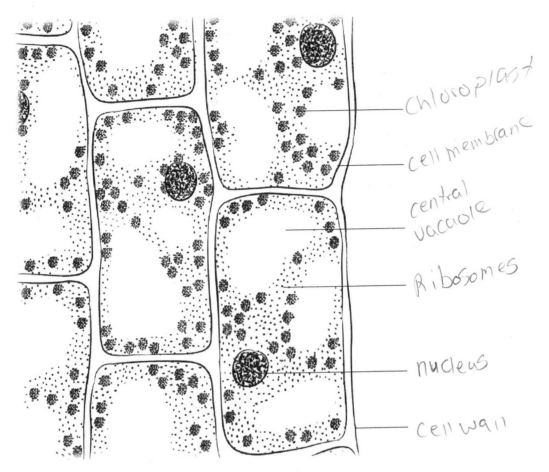

FIGURE 3-2C.   Diagram of Elodea Cells

Answer the following questions based on your observations through the microscope:

5.   Which organism(s) can carry on photosynthesis? *elodea leaf*

What **evidence** did you see that led you to this **conclusion?**

Chloroplasts

6.   Which organism has a **cell wall?** Elodea leaf

7.   Which **cell organelles** can be seen changing position in the moving cytoplasm of each organism?

Elodea   Chloroplast, ribosomes

Amoeba   ribosomes, cilia

8. Some organisms can secure food by surrounding their prey with cell extensions. This process is called **phagocytosis** (shown in **Figure 3-3**).

   Which organism(s) do you think would be able to do this? ___amobea___

   Explain your answer.

   The animal cells need food

   Plant cells use Photosynthes

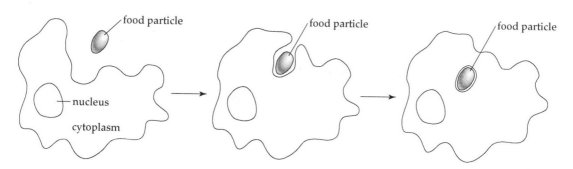

FIGURE 3-3. Phagocytosis

9. Which organism demonstrates **locomotion** (ability to move from place to place)?

   ___amobea___

Check your answers with your instructor before you continue.

ACTIVITY 3   WATER MOLECULES IN MOTION

You have just observed the movement of cytoplasm. Everywhere around us, molecules of liquids and gases are in constant, random motion.

Of course, these moving molecules are much too small to be observed directly, but we can demonstrate this activity by placing some colored powder into a drop of water.

1. Work in groups. Get the following supplies: **a slide, a cover slip, a dropper bottle of distilled water, a toothpick, and carmine powder.**

2. Place a drop of distilled water on the slide. **Carefully** add a **tiny** amount of carmine powder to the water drop. Use the **pointed tip** of the toothpick to scoop up the carmine powder.

Hint:
Imagine how much carmine powder you think you might need, and then scoop up half that amount.

Add a cover slip and observe the slide under **low power,** then **medium power,** and finally, **high power.**

3. **Describe** the activity of the carmine particles in the water droplet.

    They are not dissolved, but the particles are constantly moving

4. Since carmine particles are **not** alive, what's **causing** the carmine particles to move?

    The water molecules are bouncing off the carmine particles causing the carmine particles to keep moving

As was mentioned earlier, molecules of liquid are always in motion. This means water molecules everywhere are constantly moving and bumping into each other.

As they move, they also collide with anything floating in the water, such as the carmine particles.

5. **(Circle one answer.)** If we warmed up the carmine slide a little bit, the rate of molecular motion would be faster / slower / stay the same. **Explain** your answer.

The heat speeds them up

The process of **diffusion** occurs whenever dissolved particles move from an area of **high concentration** (more of them) to nearby areas where they are **less concentrated.** The diffusion process is similar to a bumper car ride (see **Figure 3-4**). Cars are moving quickly and often collide. The force of the collision sends the cars shooting outward into an empty area of the bumper car rink. The more an area is crowded with bumper cars, the more collisions occur.

In this way, the cars are gradually dispersed from the crowded center of the rink (an area of **high** bumper car concentration) to the emptier fringe areas (areas of **lower** bumper car concentration). This is a good model of what happens when particles diffuse in the environment (see **Figure 3-4**).

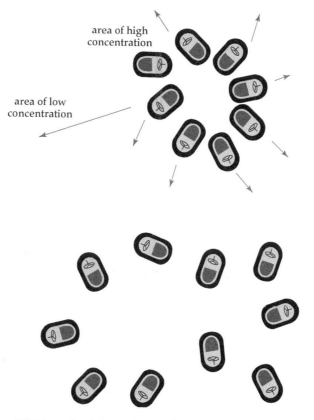

area of high
concentration

area of low
concentration

FIGURE 3-4.   Bumper Car Model of Diffusion

Diffusing molecules always move outward from an area of high concentration into areas of lower concentration. The difference in concentration levels between two locations (for example, inside the cell compared to outside the cell) is known as the **concentration gradient.** When molecules move from an area of high concentration to an area of lower concentration, we say they are moving **with the concentration gradient.**

Water molecules disperse through a **selectively permeable membrane** from an area of **high concentration** to an area of **lower concentration.** Although the mechanism is similar to that of diffusion, the process is given a different name, **osmosis.**

Solutions used by cells have two components: dissolved molecules (called **solutes**) and water (the **solvent**). The cell membrane, which controls the movement of materials in and out of the cell, can't control the movement of all molecules. **Carbon dioxide, oxygen,** and **water** are examples of molecules that move passively through the cell membrane following the concentration gradient.

We can demonstrate the processes of diffusion and osmosis in an experiment using **dialysis tubing,** an artificial membrane with many small pores. We'll be testing a solution with **three solutes: sugar, salt, and starch.**

ACTIVITY 4   DIFFUSION AND OSMOSIS

1. Work in groups. Get the following supplies: **one plastic jar, one piece of dialysis tubing, and two long pieces of thread.**

    Dialysis tubing is an example of a selectively permeable membrane. Cell membranes are also selectively permeable.

2. **Wet the piece of dialysis tubing** for a minute or two and open it as demonstrated by your instructor.

    Construct a **dialysis tubing bag. Fold about one inch** of the tubing at one end, as shown in **Figure 3-5. Tie it tightly** with the thread. The thread should be wrapped several times around the **doubled** end of the tubing. Cut off any excess string.

FIGURE 3-5. Dialysis Tubing Bag

3. Get a bottle of **stock solution.** The solution consists of **water** with a mixture of dissolved **salt** (NaCl), **sugar** (glucose), and **starch.**

Hint:

Starch tends to settle out on the bottom of the bottle. Shake the stock solution to mix it completely before measuring.

---

4. Using a graduated cylinder, measure **20 milliliters (ml)** of stock solution and pour it into your dialysis tubing bag.

    Repeat the procedures outlined in step 2 above to seal the open end of the dialysis tubing bag. Cut off any excess string.

5. **Weigh your bag** according to the following directions, using the balance in your classroom:

    ■ Make sure the **balance is set to zero.** (If you have any problems using the scale, consult your laboratory instructor.)

    ■ Weigh the bag to **the nearest tenth of a gram** (for example, **19.4 g**)

    ■ Record the **initial weight** here: _____ g

6. Fill the plastic jar about three quarters full with **distilled water.** Don't fill the jar with tap water. **Immerse** the dialysis tubing bag in the jar of distilled water.

7. **Write a hypothesis.** What do you think will happen to the **weight** of the bag during the 20 minutes it sits in the jar of distilled water?

    Hypothesis:

8. Set the jar aside for **at least 20 minutes.**

    While you're waiting, complete the Comprehension Check and then set up the controls for chemical testing in Activity 5.

    You'll return to your dialysis experiment when **at least 20 minutes** have elapsed.

✓ Comprehension Check

1. When someone uses a 10% glucose solution, what is actually in this solution? Fill in the blanks with the correct percentages (this **must** add up to 100%).

    _____ % glucose _____ % water

2. **Figure 3-6** represents a dialysis tubing bag containing dissolved salt, sugar, and starch, surrounded by distilled water.

FIGURE 3-6.   Dialysis Bag Surrounded by Distilled Water

(**Circle one answer.**)

The **highest** concentration of **salt** is **inside the bag / outside the bag.**

The **highest** concentration of **sugar** is **inside the bag / outside the bag.**

The **highest** concentration of **starch** is **inside the bag / outside the bag.**

The **highest** concentration of **water** is **inside the bag / outside the bag.**

   Check your answers with your instructor before you continue.

ACTIVITY 5   CONTROLS FOR THE DIFFUSION AND OSMOSIS EXPERIMENT

The purpose of the dialysis tubing experiment is to determine which molecules are able to cross the selectively permeable membrane of the dialysis tubing bag. To determine whether starch, salt, and sugar are able to cross the membrane, you'll perform **three** different chemical tests. Each uses an **indicator chemical that changes color** in the presence of **one** of the following molecules: **starch, salt,** or **sugar.**

a. **Iodine Test for the presence of starch.** When added to a solution, iodine will turn **black** if starch is present. If no starch is present, iodine will remain **reddish-brown** in color.

b. **Silver Nitrate Test for the presence of chloride ions.** When added to a solution that contains **chloride ions,** silver nitrate will change color from **clear** to **cloudy white.** (Remember: salt is composed of sodium ions and chloride ions.)

c. **Benedict's Test for simple sugars.** When added to a solution that contains simple sugars, such as glucose, Benedict's solution will change color from **turquoise blue** to one of the following colors: **green, yellow, orange,** or **red.** Green indicates the smallest amount of simple sugar and red indicates the highest.

> **Note:**
>
> It's important to remember that the starch, salt, and sugar are not changing in any way. It's the indicator chemicals that are changing color.

---

1. Work in groups. Get a **large beaker.** Fill it **one-third full** with **tap water** and heat it to boiling.

   While you're waiting for the water to boil, set up the materials needed for chemical testing.

2. To set up your **control** tubes, get the following supplies: **a test tube rack, a ruler, a marking pencil, a test tube holder, and three test tubes.**

   Label the test tubes salt, sugar, and starch.

3. Using the ruler, place a **line** on each test tube **1 cm from the bottom** of the tube.

   Fill each of the three test tubes to the 1-cm line with stock solution.

4. To the test tube labeled **salt,** add **one dropper full** of **silver nitrate** solution.

Caution!

Avoid getting this solution on your skin.

---

5. Shake the tube gently to mix the contents. **Observe the color** of the solution. **Record your results** in the appropriate column of **Table 3-2.**

Note:

Save all three of your control test tubes (salt, starch, and sugar) for later use.

---

6. To the test tube labeled **starch,** add **one dropper full** of **iodine** solution. Shake the tube gently to mix the contents.

   Observe the color of the solution. **Record** your results in **Table 3-2.**

7. To the test tube labeled **sugar,** add **one dropper full** of **Benedict's** solution. Shake the tube gently to mix the contents.

8. **Carefully** lower the test tube into the boiling water bath and allow it to remain for about **two minutes.**

Caution!

Hot glass looks exactly like cold glass. Use a test tube holder to remove test tubes from the boiling water.

Don't leave the test tube holder clamped on the test tube while in the boiling water. Hot metal looks exactly like cold metal!

---

9. **Using a test tube holder,** remove the tube from the boiling water bath and place it in the test tube rack.

10. **Observe the color** of the solution. **Record** your results in **Table 3-2.**

11. Form **hypotheses** for your experimental results by **circling one answer.**

    Hypotheses:

    If **starch** leaves the dialysis tubing bag, I expect that the **iodine** test results on the water **in the jar** will be **positive / negative.**

    If **sugar doesn't** leave the dialysis tubing bag, I expect that the **Benedict's** test results on the water **in the jar** will be **positive / negative.**

    If **salt** leaves the dialysis tubing bag, I expect that the **silver nitrate** test results on the water **in the jar** will be **positive / negative.**

 Check your answers with your instructor before you continue.

TABLE 3-2				
RESULTS OF CHEMICAL TESTS				
INDICATOR SOLUTION	SHOWS THE PRESENCE OF	INITIAL COLOR	FINAL COLOR	RESULTS (+ OR −)
Silver Nitrate Test Control tube				
Iodine Test Control tube				
Benedict's Test Control tube				
Silver Nitrate Test Experimental tube				
Iodine Test Experimental tube				
Benedict's Test Experimental tube				

ACTIVITY 6
COMPLETING THE DIFFUSION AND OSMOSIS EXPERIMENT

1. Remove your dialysis tubing bag from the jar of water to test for osmosis. Follow the weighing instructions in **Activity 4, step 5** to obtain the final weight of your bag.

2. Record the **final weight** here: _____ g

   Did the bag **gain** or **lose weight**? _____

   Was your hypothesis about the **weight change** correct? _____

3. Calculate the **percent change in weight** and record your answer.

   $$\text{percent change} = \frac{\text{final weight} - \text{initial weight}}{\text{initial weight}} \times 100$$

   percent weight change = _____ %

4. What do you think **moved through the membrane** to cause this result?

5. To test for **diffusion**, get **three more test tubes. These are the three experimental tubes. Label the test tubes salt, sugar, and starch.**

   Using the ruler, place a **line** on each test tube **1 cm from the bottom** of the tube.

   Fill each of the three test tubes to the 1-cm line with water from the plastic jar that held the dialysis tubing.

   Test the water for salt, sugar, and starch using the same indicator chemicals you used when setting up the controls in **Activity 5.**

   Each test tube should contain only water from the jar and the indicator chemical.

   Don't add stock solution to these test tubes!

6. **Record** the results of all three tests in **Table 3-2.**

   According to your test results, **list the molecules** that were able to **diffuse** out of the dialysis bag.

✓ Comprehension Check

1. If molecules are **expected** to move from **high concentration** to **low concentration,** why didn't **all** the molecules (salt, sugar, and starch) leave the dialysis tubing bag?

2. The ability to **separate molecules from each other** using a **selectively permeable** membrane is referred to as **dialysis.** Did the experiment you just finished illustrate this concept? **Explain** your answer.

   In your experiment with the dialysis tubing, you observed that water will move through a cell membrane from an area of high concentration to an area of lower concentration. The solution with a higher solute concentration (in our example, this was the solution inside the dialysis tubing bag) is said to be hypertonic to the solution with a lower solute concentration (the water in the beaker). You can reverse this statement and say that the water in the beaker is hypotonic (has a lower solute concentration) to the fluid inside the dialysis tubing bag. Solutions with equal solute concentrations are said to be isotonic to each other.

A student performed an experiment on **osmosis** using **human blood cells.** She exposed the cells to various solutions and observed the results through the microscope. The results of her experiment are shown in **Figure 3-7.**

3. **Draw arrows for each situation** in **Figure 3-7** showing the **direction of water movement** through the cell membrane.

a

b

c

FIGURE 3-7. Blood Cells Exposed to Various Concentrations of Extracellular Fluid

4. **(Circle one answer.)** The salt water represents a **hypertonic / isotonic / hypotonic** solution compared to the red blood cells.

5. **(Circle one answer.)** The distilled water represents a **hypertonic / isotonic / hypotonic** solution compared to the red blood cells.

6. **(Circle one answer.)** The plasma represents a **hypertonic / isotonic / hypotonic** solution compared to the red blood cells.

 **Check your answers with your instructor before you continue.**

Self Test

1. Of the cell organelles studied in this exercise, **list three** that are found in plant cells.

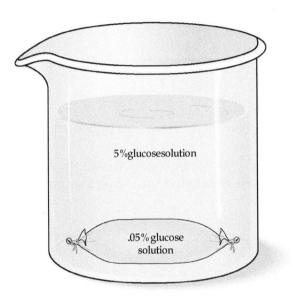

FIGURE 3-8. Movement of Water Molecules

2. In **Figure 3-8, draw arrows** showing the movement of the water molecules.

3. A patient with partial kidney failure enters the hospital. The doctor tells the patient that the kidney is responsible for filtering and removing waste molecules from body fluids. The doctor recommends use of a **kidney dialysis** machine. Using your knowledge of **diffusion, osmosis,** and **dialysis,** what do you think the dialysis machine will do for this patient?

4. You're admitted to the hospital with severe dehydration. A student nurse is directed to give you a transfusion of **blood plasma** to replace your lost body fluids. The nurse gives you a transfusion of **distilled water** by mistake. **Explain** what will happen to your **red blood cells.** In your explanation, use the following words: **cell membrane, osmosis, high concentration, and low concentration.**

5. Will a sugar cube dissolve faster in iced tea or hot tea? _____

    Explain your answer. In your explanation, use the following words: **diffusion, high concentration, low concentration, molecular movement, and collisions.**

6. You're in a restaurant that's divided into "smoking" and "nonsmoking" sections. There's a partition between the two sections, but the partition doesn't reach the ceiling. Despite the fact that you're seated in the nonsmoking section, you're bothered by the smell of cigarette smoke during your meal. Explain why the cigarette smoke was present in the nonsmoking section of the restaurant. In your explanation, use the following terms: **diffusion, molecules, higher concentration, and lower concentration.**

7. Roads are sometimes salted to melt ice. What effect does this have on plants growing along the road? **Explain** your answer.

Investigating Cellular Respiration

Objectives

After completing this exercise, you should be able to:

- explain the process of aerobic cellular respiration in plants and animals
- summarize experimental results showing that microorganisms also perform aerobic cellular respiration
- compare the rate of aerobic cellular respiration and carbon dioxide production while at rest with that during exercise
- explain the similarities and differences between aerobic respiration and anaerobic respiration and give examples
- explain the similarities and differences between ethanol fermentation and lactic acid fermentation
- apply your knowledge of aerobic and anaerobic respiration to real-life situations

CONTENT FOCUS

Cellular respiration is the process by which living organisms convert the **chemical energy** in food **(organic molecules)** to a useful form for cellular functions. This conversion process is comparable to the production of electrical energy to run machines in your home. A television can't take energy directly from coal, gasoline, or even nuclear fuel. These fuels must be converted to electricity by your community power plant before they can be used by your home appliances.

Cells must convert the energy stored in food molecules into **adenosine triphosphate (ATP),** a form of energy that can be used to run cell activities. If you read the ingredients on a candy wrapper or a box of corn flakes, however, you won't find ATP listed. The energy in the foods has to be **converted** into ATP. This conversion process is called **aerobic cellular respiration.**

During **aerobic respiration,** living organisms extract energy from the chemical bonds in food molecules (such as carbohydrates, fats, or proteins) and convert that energy into ATP.

The process of cellular respiration can be summarized with this equation:

food molecules + oxygen → carbon dioxide + *ATP energy* + water + heat

Cellular respiration isn't the same as breathing. It occurs continuously, day and night, in all living cells. If it stops, cells will quickly die. Aerobic respiration is not a single chemical reaction, but involves as many as 50 intermediate steps and the formation of a number of different compounds before ATP energy is finally produced. Specific enzymes are needed to control each step. For all living organisms, the chemical reactions of cellular respiration are amazingly similar. Many pathways are virtually identical, even between such dissimilar organisms as bacteria and human beings.

The process of aerobic cellular respiration **uses oxygen** and **produces carbon dioxide as a waste product.** For this reason, cellular respiration can't take place without gas exchange (breathing). When you **inhale, oxygen is carried by the bloodstream** to all the cells of your body. As **blood circulates** through your tissues, it **picks up the carbon dioxide produced by cellular respiration** and transports it to the **lungs** for removal. Every time you **exhale, carbon dioxide** is released.

To sum up, aerobic respiration is a process that uses oxygen to burn the food (fuel) we consume and produce useful ATP energy. The more energy your body needs, the more fuel and oxygen you must take in. It should now be clear why your breathing rate (and therefore your oxygen intake) increases when you exercise. During exercise, you also increase the amount of carbon dioxide you produce, since you burn more food molecules to produce ATP.

ACTIVITY 1   EVIDENCE THAT CARBON DIOXIDE IS RELEASED DURING CELLULAR RESPIRATION

From the following equation, you can see that organisms produce carbon dioxide gas (CO_2) during aerobic cellular respiration.

food molecules + oxygen → *carbon dioxide* ꟾ ATP energy + water + heat

Bromothymol blue is a solution that **turns green or yellow** when CO_2 is added and **turns back to blue** when CO_2 is removed. We'll use bromothymol blue as an **indicator** that will tell us if CO_2 is produced.

1. Work in groups. In your classroom, you'll find a demonstration of cellular respiration. The demonstration setup consists of a paper bag covering a rack of sealed test tubes containing **bromothymol blue and sample organisms (see Figure 4-1).**

> ### Note:
>
> **Read through the instructions COMPLETELY and record your hypotheses in Table 4-1 BEFORE LOOKING AT THE DEMONSTRATION!**

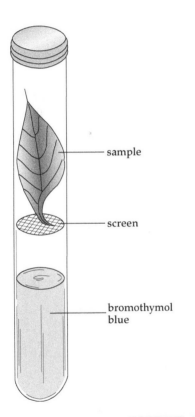

Demonstration Setup:

Tube 1	No sample
Tube 2	Seedling (young plant)
Tube 3	Plant seeds
Tube 4	Plant leaf
Tube 5	Cricket
Tube 6	Pebble

The samples are suspended on a piece of screen so that they won't fall into the bromothymol blue solution.

FIGURE 4-1. Appearance of Experimental Test Tubes

2. **Form a hypothesis for each test tube.** Which of the samples will produce CO_2 and change the color of the indicator solution from blue to yellow?

    Enter your hypotheses in the first column of **Table 4-1.**

TABLE 4-1

WHICH TUBES CONTAIN CARBON DIOXIDE?

SAMPLE	HYPOTHESIS: WILL CO_2 BE PRODUCED? (YES OR NO)	RESULTS: COLOR OF INDICATOR SOLUTION	CONCLUSION: WAS CO_2 PRODUCED? (YES OR NO)
Tube 1: No sample			
Tube 2: Seedling (young plant)			
Tube 3: Plant seeds			
Tube 4: Plant leaf			
Tube 5: Cricket			
Tube 6: Pebble			

3. Carefully lift the paper bag and **observe** the samples. **Record your observations** in the **Results column** of **Table 4-1**.

✔ Comprehension Check

1. Based on your observations, which tubes showed evidence of CO_2 production? Record your answers in the **Conclusions column of Table 4-1**.

2. Was there a control tube in this experiment? _____

   If so, which tube was the control? _____ **Explain** your answer.

3. **Based on your results,** which of the sample organisms was performing cellular respiration? **Explain** your answer.

4. **(Circle one answer.)** If one test tube contained a **dead cricket,** the amount of CO_2 in the test tube would be **more / less / exactly** the same compared to the tube with the living cricket.

5. **(Circle one answer.)** Seeds contain plant embryos. Based on the results of this experiment, the embryos in the tested seeds are probably **active / dormant (inactive).**

Check your answers with your instructor before you continue.

ACTIVITY 2   DO MICROORGANISMS PERFORM CELLULAR RESPIRATION?

Yeasts are microscopic fungi that are commercially very important. They perform aerobic cellular respiration using pathways similar to those found in larger plants and animals.

Methylene blue dye can be used as an **indicator** for cellular respiration in yeast.

Aerobic respiration releases hydrogen ions and electrons that are picked up by the methylene blue dye, gradually turning the dye **colorless.** The mitochondria of yeast cells undergoing cellular respiration will appear as a clear area surrounded by a ring of light blue cytoplasm. The nucleus may be visible as a small, darkly stained spot.

If cellular respiration isn't taking place, the mitochondria will absorb the blue dye and will not turn colorless. The cells will appear to have a large, **darkly stained** central area surrounded by the ring of light blue cytoplasm.

1. Work in groups. Get the following supplies: **slides, cover slips, yeast suspension, a pipette, and a dropper bottle of methylene blue dye.**

2. Place a **drop of yeast suspension** on a clean microscope slide. Add **one small** drop of **methylene blue** dye and place a **cover slip** over the mixture. Observe the yeast cells with the **high-power objective.**

3. Is cellular respiration occurring in any of the yeast cells on your slide? _____

   What percentage of the yeast cells are **not** undergoing cellular respiration? _____ (Make an estimate.)

   Why **isn't** cellular respiration taking place in all the cells?

4. **Draw one yeast cell** that is **undergoing cellular respiration** and **one that isn't.** Make the drawings **large and clear.** Label the **cytoplasm, nucleus (if visible),** and **mitochondria** of your yeast cells.

YEAST CELL CELL RESPIRATION OCCURRING	YEAST CELL CELL RESPIRATION ABSENT

Check your drawings with your instructor before you continue.

ACTIVITY 3   COMPARISON OF CLASSROOM AIR WITH EXHALED AIR

This experiment uses **limewater** (a very concentrated **calcium carbonate solution**) as an **indicator** solution. Limewater turns cloudy when CO_2 is added to it.

1. Work in groups. Get the following supplies: **two small beakers, one drinking straw, a pipette, and a container of limewater.**

2. **Fill** each beaker **about half full with limewater.** Label the beakers 1 and 2.

Note:
Read through the instructions COMPLETELY before you continue.

3. Do you think bubbling **room air into Beaker 1** will cause the limewater to turn cloudy? **Enter your hypothesis in Table 4-2.**

4. Do you think bubbling **your exhaled air into Beaker 2** will cause the limewater to turn cloudy? **Enter your hypothesis in Table 4-2.**

T A B L E 4 - 2 EFFECT OF BUBBLED AIR ON LIMEWATER		
	HYPOTHESIS (YES OR NO)	RESULTS
Will room air cause the limewater in Beaker 1 to turn cloudy?		
Will exhaled air cause the limewater in Beaker 2 to turn cloudy?		

5. To bubble **room air, repetitively squeeze and release** air from the **pipette** into the liquid in **Beaker 1.** Continue this process for **one minute.**

    Observe the beaker during the bubbling process and **record** your results in **Table 4-2.**

6. To bubble test your **exhaled air,** blow **very gently** through the drinking straw into the liquid in **Beaker 2.** Continue this process for **one minute.**

    Observe the beaker during the bubbling process and record your results in **Table 4-2.**

Caution!
Be careful not to suck the solution into your mouth!

7. When you've completed your experiment, empty your beakers into the **waste container.**

✔ Comprehension Check

1. In one or two sentences, **summarize the results** of this experiment.

2.   Based on your results, **what gas was bubbled** into Beaker 2? _____

3.   In reference to **Question 2,** what **process** produced this gas? _____

4.   If people in the classroom are exhaling, why do you think your Beaker 1 results were negative?

Check your answers with your instructor before you continue.

ACTIVITY 4

EFFECT OF EXERCISE ON CARBON DIOXIDE PRODUCTION

Breathing is controlled by a "breathing center" in the medulla of the brain. This center is activated by a **buildup of carbon dioxide in the blood and/or a low oxygen concentration,** such as occurs at high elevations. This experiment will compare the **amount** of CO_2 produced during exercise with that produced at rest.

   We'll use **bromothymol blue** as an **indicator** that will tell us **how much** CO_2 is produced. Recall that bromothymol blue turns **green or yellow** when CO_2 is **added** and returns to **blue** when CO_2 is **removed.**

1.   Work in groups. Get the following supplies: **three 400-ml beakers, bromothymol blue solution, ammonia solution, a plastic bag, a coupler tube, a mouthpiece, a rubber band, and an air stone assembly.**

2.   Fill each beaker **with 200 ml of bromothymol blue indicator solution. Label** the beakers **1, 2, and 3.**

3.   **Attach the coupler tube** to the plastic bag and hold the tube in place with the rubber bands (as shown in **Figure 4-2**).

Testing Room Air

1.   Fill the plastic bag with **room air** by pulling it through the air ("rounding out" the bag).

   Quickly plug the bag with the **air stone assembly. Be careful not to let the air escape.**

2. Place the **air stone** into the indicator solution in **Beaker 1** (see **Figure 4-2**).

    Squeeze the air out of the plastic bag so that it bubbles into the beaker.

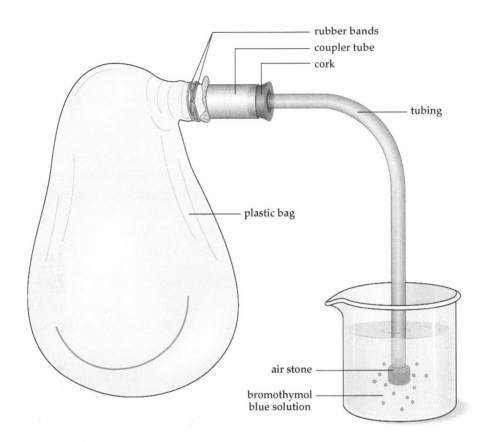

rubber bands
coupler tube
cork
tubing
plastic bag
air stone
bromothymol
blue solution

FIGURE 4-2. Apparatus to Bubble Captured Air into Bromothymol Blue Solution

Ammonia Treatment

1. Using the dropper bottle, add **ammonia** to **the beaker one drop at a time** and **gently** stir after adding each drop.

    Be sure to **keep an accurate count** of the number of drops added.

2. Continue adding ammonia until the indicator solution changes to a **light green color and remains green for 30 seconds.**

3. **Record** the number of drops you added in **Table 4-3**.

    The amount of ammonia needed to change the indicator solution from **blue to green** will be used as an estimate of **the amount of CO_2 present** in the room air.

TABLE 4-3 EFFECT OF EXERCISE ON CO_2 PRODUCTION		
SAMPLE	COLOR AFTER BUBBLING	DROPS OF AMMONIA USED
Room air (Beaker 1)		
Exhaled air at rest (Beaker 2)		
Exhaled air after exercise (Beaker 3)		

Testing Exhaled Air before Exercise (at Rest)

1. Secure the mouthpiece onto the plastic bag and fill the bag with exhaled air, crimping the end of the bag with your hand to prevent the air from escaping. **Quickly** plug the bag with the **air stone assembly,** being careful not to lose the air.

2. Place the air stone into the indicator solution in **Beaker 2.** Squeeze the air out of the plastic bag so that it bubbles into the beaker. Keep squeezing until the plastic bag is empty.

    Add ammonia one drop at a time until the indicator solution changes to a **light green color and remains green for 30 seconds. Record** the number of drops you added in **Table 4-3.**

Testing Exhaled Air after Exercise

1. Secure the mouthpiece on the plastic bag. Set the bag close at hand and **exercise vigorously for two minutes.** You may jog around the room, do jumping jacks, or step rapidly up and down on an aerobic step.

    Fill the plastic bag with exhaled air **immediately** and plug the bag with the **air stone assembly.**

2. Bubble the exhaled air into **Beaker 3** as you did when testing the exhaled air before exercise. **Add ammonia** until the indicator solution changes to a **light green color and remains green for 30 seconds. Record** the number of drops you added in **Table 4-3.**

✔ Comprehension Check

1. What happened to the amount of CO_2 in your exhaled air after exercise? **Explain** your answer.

2. How do you think it would affect the amount of CO_2 produced if you exercised for ten minutes instead of only two minutes? **Explain** your answer.

3. When you exercise, your heart beats faster and the blood carries more oxygen to your muscles. What do muscle cells do with this oxygen?

Check your answers with your instructor before you continue.

ACTIVITY 5   ALTERNATIVES TO AEROBIC RESPIRATION

You've demonstrated that cells can carry out aerobic respiration when oxygen is available, but they can also produce ATP energy by an anaerobic process called **fermentation** when oxygen is in short supply. The fermentation pathway can produce a variety of end products (refer to **Figure 4-3**).

In most plant cells and in yeasts, pyruvic acid is converted to ethanol (drinking alcohol) and carbon dioxide. Alcohol fermentation by yeast cells is used commercially to manufacture baked goods and alcoholic beverages.

Alcohol fermentation is summarized with the following equation:

food molecules → pyruvic acid → carbon dioxide + ATP energy + *ethanol* + heat

Another method of anaerobic respiration is called **lactic acid fermentation.** During lactic acid fermentation, the pyruvic acid molecules from glycolysis are converted into **lactic acid.**

Some bacteria specialize in lactic acid fermentation. The lactic acid produced by bacteria causes milk to sour. A similar process occurs when cabbage ferments to form sauerkraut. The dairy industry uses lactic acid fermentation by bacteria and fungi to produce sour cream, cheese, and yogurt. Your muscle cells may use lactic acid fermentation when oxygen demands can't keep up with supply (such as during vigorous exercise).

The equation for lactic acid fermentation is

food molecules → pyruvic acid → ATP energy + *lactic acid* + heat

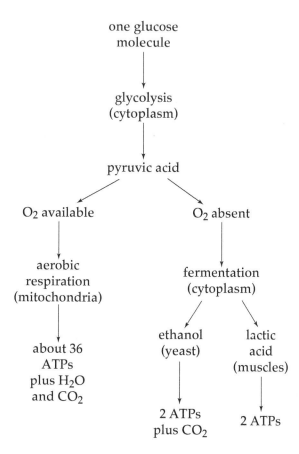

FIGURE 4-3. Steps of Aerobic and Anaerobic Cellular Respiration

✔ Comprehension Check

The steps that occur during aerobic and anaerobic respiration are summarized in **Figure 4-3**. Refer to **Figure 4-3** and answer the following:

1. If **oxygen is available,** how many ATPs can a cell produce from **one** glucose molecule? _____

2. How many food molecules would a cell need to process to obtain the same amount of energy from **anaerobic** respiration? _____

3. Name the cell location where **glycolysis** (the first step of cell respiration) occurs.

    _____

4. Name the part of a cell in which **aerobic** respiration pathways occur. _____

5. Name the part of a cell in which **anaerobic** respiration pathways occur. _____

Check your answers with your instructor before you continue.

ACTIVITY 6 ANAEROBIC RESPIRATION

> **Note:**
>
> **Read through the instructions COMPLETELY before you begin.**

1. Work in groups. Get the following supplies: **a pair of safety glasses, a tablespoon, two 400-ml beakers, a container of limewater, and a jar of sucrose.**

   In your classroom, you'll find two **flasks** on a **warming plate.** The flasks contain **water** and **yeast cells.**

   Each flask is closed with a stopper holding two pieces of tubing (see **Figure 4-4**). The longer piece of tubing is connected to an **air stone.**

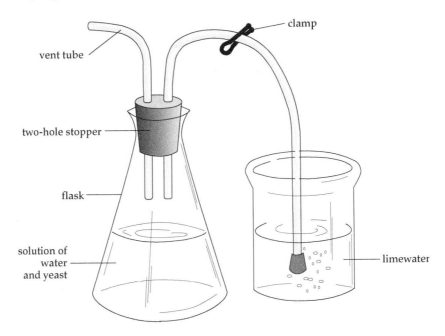

FIGURE 4-4. Setup for a Fermentation Experiment

2. Fill both beakers **half full** with **limewater**. Place the beakers on the table next to the warming plate. Insert the air stone from **Flask A into the first beaker** and from **Flask B into the second beaker.**

Caution!
Eye protection is required for the following procedures!

3. Wearing eye protection, gently remove the stopper from **Flask B and add one tablespoon of sucrose.**

    Replace the stopper. **Record the time** when you added the sucrose. _____

4. **Immediately** after adding the sucrose, remove the clamp from the air stone tube and **switch it to the vent tube. Switch the clamps on Flask A and Flask B.**

5. Check on your experiment every **five minutes** for the next **15 minutes. Record** the results for **each five-minute period** in Table 4-4.

TABLE 4-4

OBSERVATIONS OF YEAST FLASKS

OBSERVATIONS	EXPERIMENTAL FLASK A (WATER AND YEAST)			EXPERIMENTAL FLASK B (WATER, YEAST, AND SUCROSE)		
	5 MIN.	10 MIN.	15 MIN.	5 MIN.	10 MIN.	15 MIN.
Color of liquid in yeast flask						
Color of lime water in beaker						
Presence of bubbles in yeast flask						

6. Did the results for Flasks A and B differ? **Explain** your answer by mentioning facts collected during this experiment.

7. What gas formed the bubbles observed in the flask(s)? _____

    Explain your answer.

Caution!

Eye protection is required for the following procedures!

8. Wearing eye protection, carefully remove the stopper from **Flask A** and sniff the released air. **Repeat** the process with **Flask B. Describe** the results.

✓ Comprehension Check

1. A yeast cell undergoing fermentation produced **six ATP molecules.** How many glucose molecules were used?_____ **Explain** your answer.

2. If the same cell was undergoing aerobic cellular respiration instead of fermentation, but using the same number of glucose molecules, how many ATP molecules would be produced?_____ **Explain** your answer.

 Check your answers with your instructor before you continue.

Note:

Lactic acid fermentation can only satisfy the body's short-term energy needs. If the oxygen shortage continues too long, lactic acid accumulates in your muscle tissue, lowering the pH. Eventually, this acidic environment inhibits the activity of enzymes that are needed for muscle function. Muscle fibers lose their ability to contract, a process called muscle fatigue.

The accumulated lactic acid can damage muscle proteins, producing the soreness and pain often associated with lengthy, strenuous exercise.

When oxygen is available again, the lactic acid is converted back to pyruvic acid and aerobic respiration can continue as before.

Self Test

1. A rock covered with green spots is placed on a screen in a test tube above some bromothymol blue solution; after several hours, the solution remains blue. What can you **conclude** from this experiment?

2. You and several other students observe that dead animals along the roadside often increase in size several days after they have been killed. In the spirit of scientific investigation, you collect the gas from inside one of these dead animals and bubble it through limewater. The limewater turns cloudy.

    a. What does the cloudy result indicate?

    b. If the animal is dead, how are these results possible? **Explain your answer in detail.**

3. Which of the following produces the most energy in the form of ATP?

    a. aerobic respiration
    b. anaerobic respiration
    c. alcohol fermentation
    d. all of the above produce the same amount of ATP energy, but through different chemical reactions
    e. none of the above produces ATP

4. Carbon dioxide is passed into a solution of bromothymol blue indicator until the solution turns yellow. A sprig of elodea is then placed in this yellow solution. After a few hours in the sunlight, the yellow solution turns blue again. **Suggest an explanation** for the color changes observed in the bromothymol blue.

For questions 5 through 8, use the following three answers. An answer may be used once, more than once, or not at all. For each question, explain your answer.

    a. process occurs only under aerobic conditions
    b. process occurs only under anaerobic conditions
    c. process occurs under both aerobic and anaerobic conditions

5.   _____ Manufacturing beer or wine.

    Explanation:

6.   _____ ATP production.

    Explanation:

7.   _____ Accumulation of lactic acid in a runner's muscles causes her to drop out of the race.

    Explanation:

8.   _____ Production of 36 ATPs, water, and carbon dioxide from one glucose molecule.

    Explanation:

9. If an elephant's cells were to lose their ability to perform aerobic cellular respiration, which of the following would most likely happen?

    a. the cells would switch over to anaerobic respiration and the elephant would be fine

    b. the cells would switch over to lactic acid fermentation and the elephant would survive, but would have achy muscles

    c. the cells would switch over to alcohol fermentation and the elephant would survive but would be in a drunken stupor

    d. the cells would die because the elephant's energy needs can't be met using anaerobic metabolic pathways

    e. the cells could live on stored energy for an extended period and cellular respiration wouldn't be necessary for several years

    Explain your answer.

Enzyme Activity

Objectives

After completing this exercise, you should be able to:

- discuss the basics of enzyme function in cells
- explain the relationship between the three-dimensional structure of proteins and enzyme function
- describe the effects of various environmental conditions on protein denaturation
- explain the activity of digestive enzymes in food vacuoles.

CONTENT FOCUS

Within living systems, chemical reactions require specific **enzymes** to assist and speed the rate of the reactions (these "helper" molecules are known as **catalysts).** Enzymes are **proteins,** and they are very specific, each working with only one or very few chemical compounds. In addition, different enzymes work best under different environmental conditions.

In order for a protein to function correctly, it must be folded into a specific three-dimensional shape. In enzyme molecules, folding creates a region called the **active site,** where molecules can bind to the enzyme and a reaction can take place.

If environmental factors (such as pH, temperature, or salinity) lead to changes in the specific three-dimensional shape of the enzyme's active site, the enzyme may not function correctly. When an enzyme's folding is altered, the enzyme is said to be **denatured.**

The following activities will demonstrate how enzymes work and the effect of some environmental conditions on protein structure.

ACTIVITY 1   DEMONSTRATION OF ENZYME ACTIVITY

Amylase is an enzyme that **hydrolyzes starch** (a polysaccharide). In humans, amylase is present in the mouth and the small intestine.

To demonstrate the action of amylase on starch molecules, two indicator tests will be used.

Iodine is a chemical indicator that changes color (turns dark blue or black) in the presence of **starch.** In the presence of monosaccharides or disaccharides, iodine retains its normal reddish-brown color.

Benedict's solution is a chemical indicator for the presence of **simple sugars** (monosaccharides and some disaccharides). In the presence of simple sugars, Benedict's solution changes color from turquoise blue to one of the following colors: green, yellow, orange, or red. Green represents the smallest amount of simple sugar and red the highest. Benedict's test differs from the iodine test in that the reaction only takes place when the solution is heated.

1. Work in groups. Get the following supplies: **four large test tubes, two glass stirring rods, a test-tube rack, a test-tube holder, two graduated 10-ml pipettes with manual dispenser, a hot plate, a dropper bottle of iodine solution, a dropper bottle of Benedict's solution, a stock bottle of starch solution, a stock bottle of amylase solution, and a large beaker.**

2. Place the large **beaker,** about **one-third filled** with tap water, onto the **hot plate.**

    Set the temperature control to high until the water boils.

3. **Label the test tubes** as follows: **B-1, B-2, I-1** and **I-2.**

    Make sure the bottle of starch solution is **well mixed.**

    Using a **10-ml graduated pipette,** transfer **15 ml of starch solution** from the stock container into **each** of the large test tubes.

4. To the test tube labeled **B-1,** add **two droppers full of Benedict's solution.**

    What color is the liquid in **tube B-1?** _Light blue_

5. The chemical indicator in Benedict's solution **will only react with simple sugars when it is heated up.**

    Using a **test-tube holder, carefully** place the **tube B-1** into the boiling water for **two minutes.**

Caution!

Hot glass looks exactly like cold glass! Use the test-tube holder to remove test tubes from the boiling water!

Don't leave the test-tube holder clamped to the test tube while in the boiling water. Hot metal looks exactly like cold metal!

6. **Using the test-tube holder, remove** the test tube from the boiling water and **observe** the color of the Benedict's solution.

   Don't dispose of tube B-1.

   Was your Benedict's test result for tube **B-1 positive or negative (+/−)?**
   _____−_____

7. To the test tube labeled **I-1,** add two droppers of iodine solution.

   Don't add indicator solution to test tube **I-2.**

8. Observe the color of the indicator solution in **tube I-1.**

   Was your iodine test result **positive or negative (+/−)?** ____+____

   Don't dispose of tube I-1.

9. Using a **clean 10-ml graduated pipette,** add **15 ml of amylase solution** to tubes **B-2 and I-2.**

   Record the time when you add the amylase to the starch solution: _1500_

   Place a **glass stirring rod** into each tube: **B-2** and **I-2.**

   Set tubes B-2 and I-2 aside for **30 minutes.**

   Every five minutes, stir the contents of each tube with the rod.

10. After the 30-minute interval, add **two droppers full of Benedict's solution to tube B-2.**

    Add two droppers full of iodine solution to tube I-2.

    Shake the test tubes to mix the contents.

11. Observe the color of the indicator solution in **tube I-2.**

    Was your iodine test result **positive or negative (+/−)?** ____+____

12. If the **iodine test is negative in tube I-2,** what would this tell you about the digestion of starch by amylase?

Iodine is not present.
It is working Properly

13. Imagine that you get a **positive result for the iodine test in tube I-2.** You form the **following hypothesis** about the reason for the positive test result:

The amylase enzyme was inactive and therefore did not digest any starch.

Suggest a method by which this hypothesis could be tested.

Try another test, and this time put a sample
of starch in the tube to show it is present.
do another test and not have any starch to
Show its not present, and a control.

Check your answers with your instructor before you continue.

14. **Using the test-tube holder,** place **tube B-2** in the boiling water for **two minutes.**

Remove the test tube from the boiling water and **observe** the color of the Benedict's solution.

Was your Benedict's test result **positive or negative (+ / −)?** _____

15. Do the test results **support your hypothesis? Explain your answer.**

no because the amolazy did not
Ionize the starch.

✓ Comprehension Check

1. If your Benedict's test result was positive, where did the sugar come from? **Explain your answer.**

2. What was the purpose of the indicator tests conducted on tubes **I-1 and B-1?**

3. Why were tubes **I-1 and B-1** saved until the end of the experiment?

Check your answers with your instructor before you continue.

ACTIVITY 2   ENZYME ACTIVITY IN FOOD VACUOLES

Paramecium is a small one-celled organism found in freshwater. By means of rapid swimming motions, it funnels water containing bits of organic matter, such as yeast and bacteria, into a groove on its surface. From this groove, the materials enter the *Paramecium* through a "mouth" and are taken up by small organelles called **food vacuoles.**

The organic matter constitutes the *Paramecium's* food. Enzymes are released into the vacuoles and the food within them is digested.

1. Work in groups. Get the following supplies: **slides, coverslips, a compound microscope, a bottle of methyl cellulose (the bottle may also be labeled Protoslo®), toothpicks, a** *Paramecium* **culture, and a solution of yeast stained with congo red.**

2. Make a wet mount using the *Paramecium* culture and add **one drop** of **red-stained yeast.**

    Add a **small** drop of the methyl cellulose to your slide, **gently** stir with a toothpick, and add a coverslip.

3. Using the **scanning lens** of the microscope, locate one or more organisms.

    Switch to a higher power and locate the food vacuoles within the cell (refer to **Figure 5-1**).

    What is the initial color of the food vacuoles? *Clear*

    Observe the food vacuoles in various individuals every few minutes for **fifteen minutes.**

    Describe the **color changes** that occur within the vacuoles.

    The vacuoles have an acidic
    Ph, so they are turing blue

4. Congo red is an **indicator chemical** that turns **blue** under **acidic** conditions and **red** under **basic** conditions.

    What does this tell you about the pH inside a food vacuole?

    The PH inside a food vacuole
    is acidic

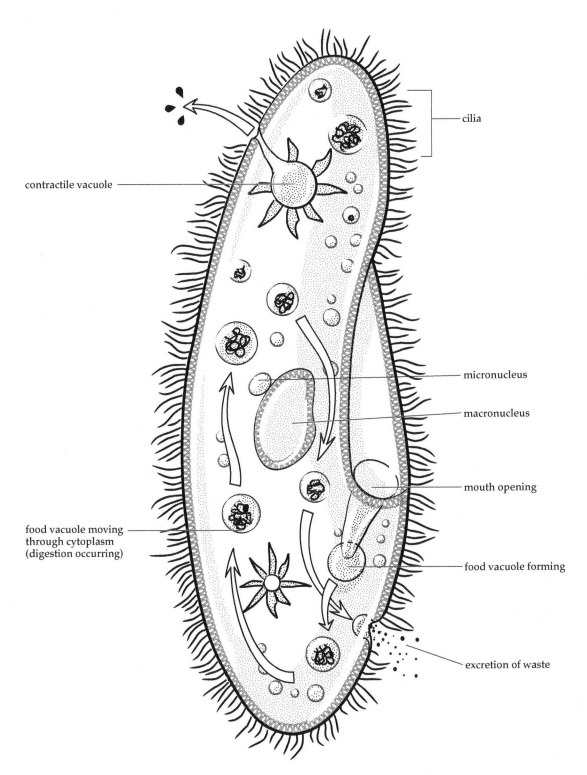

FIGURE 5-1. *Paramecium* Food Vacuole Formation

✓ Comprehension Check

1. **Pepsin** is a human digestive enzyme found in the stomach. If pepsin were present in *Paramecium*, do you think this enzyme could be active within a food vacuole? **Explain your answer.**

2. If a **base** (**alkaline solution**) was added to the yeast suspension before the *Paramecia* were fed, how might this affect the **color change** within the food vacuoles? **Explain your answer.**

3. If you ground up a "Tums" tablet to a fine powder and added it to the yeast suspension before the *Paramecia* were fed, how might this affect the **efficiency of digestion within the food vacuoles?**

Check your answers with your instructor before you continue.

ACTIVITY 3

EFFECT OF ENVIRONMENTAL CONDITIONS ON PROTEIN STRUCTURE

During this laboratory exercise, you'll be examining the effect of various environmental factors (temperature, salinity, and pH) on the folded structure of proteins. You will use chicken eggs, since they are among the few cells that are large enough to be observed without special equipment.

1. Work in groups. Get the following supplies: **five chicken eggs, five bowls, stock bottles of the following solutions: pH 11, pH 3, and 25% saline, a large beaker, clear plastic wrap, and a hot plate.**

2. Put about **400 ml of tap water** into the large beaker and set the hot plate on **high.** Heat the water to boiling.

3. **Form hypotheses** about the effect of each set of environmental conditions **(pH 11, pH 3, 25% saline, boiling, and room temperature)** on the yolk and the white of an egg.

   For example: no change will occur in the yolk or the white, the proteins in the egg yolk will denature, the proteins in the egg white will denature, the egg white will denature but not the yolk, etc.

   Record your hypotheses in **Table 5-1.**

TABLE 5-1		
RESULTS OF EGG EXPERIMENTS		
ENVIRONMENTAL CONDITIONS	HYPOTHESES	OBSERVATIONS OF THE EGG WHITE AND YOLK
pH 11		
pH 3		
25% saline solution		
Boiling water		
Room temperature		

4. Label **three** of the bowls as follows: **pH 11, pH 3, and 25% saline.**

    Carefully, without breaking the yolk, crack an egg into each of the three labeled bowls.

Caution!

Be careful not to spill the pH solutions on your skin. Rinse your hands or eyes thoroughly with water if contact occurs.

---

5. To the bowl labeled **pH 11,** add **100 ml of pH 11** solution.

    To the bowl labeled **pH 3,** add **100 ml of pH 3** solution.

    To the bowl labeled **25% saline,** add **100 ml of 25% saline** solution.

    Set the bowls aside undisturbed for **30 minutes.**

6. **Carefully, without breaking the yolk,** crack an egg into each of the two remaining bowls.

    To minimize odors, loosely cover the **pH 11** bowl with a sheet of clear plastic wrap.

    Set one of the bowls aside undisturbed for **30 minutes.**

7. Gently tip the egg out of the other bowl and into the beaker of boiling water.

    Let the water return to a full boil and boil the egg for **five minutes.**

    Turn the hot plate off and let the water cool to room temperature.

8. At the end of the 30-minute period, observe and record your observations for the yolk and white of each egg in **Table 5-1.**

Note:

Don't dispose of any of the eggs until you have answered the remaining questions.

9. Did any of the environmental conditions damage (**denature**) the proteins in the egg? **Explain your answer.**

10. Test the white part of the **"room temperature"** egg with pH paper.

   What is the approximate pH of the egg white? _____

   Is there a relationship between the pH results from the **room temperature** egg and your observations of the eggs exposed to **pH 3 and pH 11? Explain your answer.**

11. Was there a **control** in this experiment? If so, what was the control?

12. Why is a control desirable in an experiment of this type?

✓ Comprehension Check

1. Stomach conditions are quite acidic. Do cells of the stomach lining need protection from these acidic conditions? Why or why not?

2. Acid rain in the Northeastern United States can have a pH as low as 4.5 in some areas. How might this explain decreased survival among fish and amphibian eggs in these regions?

3. With reference to Question #2 above, why might the effect on bird eggs be **less serious** than that experienced by fish and amphibian eggs?

Check your answers with your instructor before you continue.

Self Test

1. Based on the results of your experiments, why is it considered dangerous to have an abnormally high fever?

2. Body fluids contain buffers whose function is to minimize changes in pH levels. Why is a buffering system beneficial for optimal enzyme function?

The following are the results of an experiment that examined the effect of pH on the activity of an enzyme isolated from the human digestive tract.

Tube #	pH	\multicolumn{12}{c}{Elapsed Time (minutes)}											
		2	4	6	8	10	12	14	16	18	20	22	24
1	2.0	–	–	–	–	–	–	–	–	–	–	–	–
2	3.0	–	–	–	–	–	–	–	–	–	–	–	–
3	3.3	–	–	–	+	+	+	+	+	+	+	+	+
4	3.5	+	+	+	+	+	+	+	+	+	+	+	+
5	3.7	–	–	–	–	–	–	+	+	+	+	+	+
6	4.0	–	–	–	–	–	–	–	–	–	–	–	–

3. The optimal pH for this enzyme is _____.

4. In a few sentences, summarize the experimental results for **pH 3.7.**

5. Would you characterize this enzyme as being **general or specific** in regard to its pH requirements? **Explain your answer.**

6. Enzymes regulate the production of the pigment melanin in the fur of Siamese cats and Himalayan rabbits. At normal body temperatures, the rabbits produce white fur, but their paws, ear tips, and noses have black fur. On the basis of your understanding of the effect of environmental conditions on enzyme activity, suggest one possible explanation for the difference in fur color on various parts of the body.

Food Analysis and Choices for Good Health

Objectives

After completing this exercise, you should be able to:

- test an unknown sample for the presence of organic molecules such as sugars, starch, protein, and lipids using indicator chemicals
- relate the nutrient content of a food to its original function in plants and animals
- discuss the benefits and drawbacks of using carbohydrates versus lipids as energy storage molecules for embryos
- use the Federal Dietary Guidelines to analyze the nutritional value of your meals

CONTENT FOCUS

All living organisms are composed of various types of **organic molecules** such as carbohydrates, lipids, proteins, and nucleic acids that make up their body tissues. Each of these four types of organic molecules (carbon-based compounds) has a different function in the body. The foods we eat are obtained from plants and animals. These foods are needed not only as a source of energy but also because they provide many essential nutrients. Just as different parts of your body have different structures to support their functions, the same is true of other animals and plants. The **types** of **organic chemicals present** in different body tissues are related to the **function** of those tissues.

In this exercise, you'll be analyzing various food samples to determine which organic compounds are present. Carbohydrates, lipids, proteins, and nucleic acids **can't be seen directly. Indicator chemicals** can show you if these organic compounds are present in a food sample by **changing color.** Each type of indicator chemical is specific for one type of organic compound.

Analyzing foods for organic compounds using indicator chemicals can present some problems. Often, the **food itself has colors** that may interfere with the test results. It's easier to determine if an appropriate color change has occurred if you have a **standard for comparison.** Therefore, we must first establish a set of **positive and negative standards (controls)** for comparison in later experiments.

ACTIVITY 1	POSITIVE AND NEGATIVE INDICATOR TESTS

1. Work in groups. Get the following supplies: **a metric ruler, a marking pencil, six test tubes, a test tube rack, a test tube holder, and a large beaker.**

2. Using the ruler, place a **line** on each test tube **1 cm from the bottom** of the tube. **Number each test tube 1 through 6.** Place the marked test tubes in the test tube rack.

Benedict's Test for Simple Sugars

1. The chemical indicator in Benedict's solution reacts with **monosaccharides and some disaccharides;** however, for a reaction to occur you have to heat the Benedict's solution. Place the large beaker, half filled with tap water, on a hot plate. Set the temperature control to high until the water boils.

2. Fill **test tube 1** up to the line with **glucose** solution.

> **Note:**
>
> **Glucose is a monosaccharide, so it can be used as a positive control for Benedict's test.**

Fill **test tube 2** up to the line with **distilled water.**

3. Fill the dropper with Benedict's solution. What color is Benedict's solution?

_____

4. To **test tubes 1 and 2,** add **one dropper full of Benedict's solution. Shake** the test tubes to mix the solution.

5. **Using a test tube holder, carefully** place the tubes into the boiling water for two to five minutes.

> **Caution!**
>
> **Hot glass looks exactly like cold glass! Use the test tube holder to remove test tubes from the boiling water!**
>
> **Don't leave the test tube holder clamped on the test tube while in the boiling water. Hot metal looks exactly like cold metal!**

6. **Using the test tube holder,** remove the tubes and **observe** the color in each test tube.

   If you see a color change ranging from **green through yellow, orange, or red,** this is a positive test for simple sugars.

   If the color in the tube **hasn't changed,** this is a **negative** test for simple sugars.

   What happened to the **color** in **tube 1?**_____

   What happened to the **color** in **tube 2?**_____

> ### Note:
> **Save tubes 1 and 2 for use in later experiments.**

✓ Comprehension Check

1. Which tube showed a positive test for glucose?_____

2. Based on the **results of Benedict's test, glucose** could be correctly identified as belonging to which of the following groups?

   a. lipids   d. simple sugars
   b. starches   e. choices b and d are both correct
   c. proteins

Iodine Test for Starch

1. Fill test tube 3 up to the line with starch solution. Fill test tube 4 up to the line with distilled water.

2. Fill the **dropper with iodine.** What color is iodine?_____

3. To **both tubes,** add one **dropper full of iodine. Shake** the test tubes to mix the solution.

4. **Observe** the color in each test tube.

   What color is the liquid in **test tube 3?** _____

   What color is the liquid in **test tube 4?** _____

5. Which tube showed a positive test for starch?_____

6. Place **one small drop of iodine** in the box.

   Based on the results of the iodine test, what can you **conclude about the paper** used for this laboratory manual?

> **Note:**
> Save tubes 3 and 4 for use in later experiments.

Biuret Test for Protein

1. Fill **test tube 5** up to the line with **albumin (egg white)** solution.

> **Note:**
> Egg white is mostly protein, so it can be used as a positive control for the Biuret test.

   Fill **test tube 6** up to the line with **distilled water**.

2. Fill the dropper with **Biuret solution.** What color is Biuret solution?
   _____

3. To **tubes 5 and 6,** add one dropper full of **Biuret solution. Shake** the tubes to mix the solution.

4. **Observe** the color in each test tube.

   What happened to the color in **tube 5?**

   What happened to the color in **tube 6?**

5. Which tube showed a **positive** test for **protein?** _____

> **Note:**
> Save tubes 5 and 6 for use in later experiments.

Sudan IV Test for Lipids

1. Get the following supplies: **one piece of filter paper, forceps, a petri dish, and dropper bottles of vegetable oil, distilled water, and Sudan IV stain.**

2. Place the filter paper on a **clean** piece of paper (**not** on the laboratory counter, which may be dirty).

   Using a **pencil** (**not** the wax pencil), draw **two** circles about the size of a dime, spaced **apart** on the filter paper. Label the first circle **"oil"** and the second circle **"water."**

3. Place **one drop of oil** into the circle labeled **"oil"** and **one drop of distilled water** into the circle labeled **"water."**

4. **Set the filter paper aside to dry.** If needed, use a hair dryer **on low power** to evaporate excess liquid.

5. When the filter paper is **completely dry,** place it in the **petri dish** and cover the paper with **Sudan IV solution.** Let the paper soak in the stain for **three minutes.**

6. While the paper is soaking, get a **glass bowl** from the **supply area** and **fill it** with **distilled water.**

7. When three minutes are up, place the filter paper into the bowl of distilled water. **Rinse gently** for **one minute.**

   Remove the filter paper from the water with the forceps and **observe the color of the two circles.**

8. **A dark red spot** indicates a **positive** test for lipids.

   A **pale pink** color, **no different from the rest of the paper,** should be considered **negative.**

9. **Record** the results of your test (positive or negative) below. Enter a " + " if the test was **positive** and a " − " if the test was **negative.**

   "oil" circle _____   "water" circle _____

Note:
Save your test paper for use in later experiments.

☑ Comprehension Check

Using the results from the tests you've just completed, fill in **Table 6-1.**

TABLE 6-1 INDICATOR TESTS			
INDICATOR TEST	TESTS FOR	NEGATIVE RESULT (COLOR)	POSITIVE RESULT (COLOR)
			green through red
	starch		
Sudan IV test			
Biuret test			
		light blue	

Check your answers with your instructor before you continue.

ACTIVITY 2   TESTING FOOD SAMPLES

Have you ever wondered what's really in the foods you choose? Is a food high in fat or protein? Is it a good source of carbohydrates? The answers to these questions become clearer when food samples are analyzed to determine which organic compounds are present.

In the next activity, you'll analyze several commonly eaten foods and determine their nutrient content. Solid food samples have been blended into liquid, for easier testing.

1. For each of the foods listed in **Table 6-2,** form a **hypothesis** about which organic compounds it will contain.

   Record your hypotheses in **Table 6-2.**

   For each hypothesis, enter a " **+** " if you think the test will be **positive** and a " **−** " if you think the test will be **negative.**

TABLE 6-2 HYPOTHESES FOR FOOD EXPERIMENTS				
FOOD	SIMPLE SUGARS	STARCH	PROTEIN	LIPIDS
Lettuce				
Hamburger				
Tuna				
Milk				
Refried beans				
Peanut butter				

2. Work in groups. Your instructor will assign several foods for your group to analyze. Your group will **share its data** with other groups that are analyzing different foods.

3. For each food sample to be analyzed, get **three clean test tubes.** You'll also need **one piece of filter paper.**

4. Following the **test procedures** exactly as you did in Activity 1, analyze each food sample for the presence of **simple sugars, starch, protein, and lipids.**

5. **Compare** your test results with the **controls** you **saved from Activity 1.**

6. **Record** the results of your tests (positive or negative) in **Table 6-3** and also on the **master chart at the front of the room.**

   Enter a " + " if the test was **positive** and a " − " if the test was **negative.**

7. **Complete Table 6-3** by entering the results from the **master chart** at the front of the room.

TABLE 6-3 RESULTS OF FOOD EXPERIMENTS				
FOOD	SIMPLE SUGARS	STARCH	PROTEIN	LIPIDS
Lettuce				
Hamburger				
Tuna				
Milk				
Refried beans				
Peanut butter				

✓ Comprehension Check

1. List the foods tested that contain **all three** of the following: **carbohydrates, proteins, and lipids.**

    It's possible to relate the nutrient content of a food to its original function in plants and animals. Using the information provided below, **explain** why certain nutrients occur in high levels in particular foods.

2. This food comes from the leaf of a plant. Leaf cells perform photosynthesis.

    Which of the six foods is being described?_____

    (Circle one answer.) If photosynthesis is taking place, I would predict a positive indicator test for **simple sugar / lipid / protein.**

3. Which of the foods tested were **positive** for **lipids?**

4. Which of the foods tested use **lipids** as an energy source for **young animals?**

5. **Muscle tissue** is composed of **interwoven protein fibers.** Using this information, **name two foods** made of **muscle** tissue that gave you a **positive** test for **protein.**

6. The hamburger and tuna tested **negative** for starch. **Explain** why foods of this type would **not** contain starch.

Check your answers with your instructor before you continue.

ACTIVITY 3   AN EMBRYO IN A PEANUT SEED

A seed contains a **plant embryo** packaged with its **food supply.** The embryo and stored food are protected by a tough seed coat. A developing embryo has the beginnings of a stem, tiny leaves, and a root. You can see the tiny stem, leaf, and root structures of an **embryonic peanut plant** if you carefully pull the two peanut halves apart.

The nutrients in seeds are not only used by germinating plant embryos but are also consumed by animals, including humans.

1. In your classroom, you'll find a demonstration of a **separated peanut seed** that can be viewed with a dissecting microscope.

2. Look for the following structures and **label** them on the diagram in **Figure 6-1: root, stem, leaves, and stored food supply.**

FIGURE 6-1. Peanut Seed

✓ Comprehension Check

1. Food may be stored for the use of an adult plant or as energy for the development of an embryo. The stored food is packaged with an embryo into a structure called a **seed.**

   List **three commonly eaten foods** produced by plants or animals that provide stored energy for embryos:

   _____   _____   _____

2. Which of the foods tested in Activity 2 use **lipids** as an energy source for a **developing embryo?**

3. **(Circle all correct answers.)** The following probably have high lipid levels:

   a. chicken egg   c. sunflower seed   e. coconut

   b. carrot   d. spinach   f. almond

   Explain your answer.

Check your answers with your instructor before you continue.

ACTIVITY 4   DIETARY INTAKE AND GOOD HEALTH

When you eat a meal, you're probably consuming a combination of carbohydrates, proteins, and lipids. Imagine this situation. You've been stuck in classes all morning (just grabbed a bag of peanuts from the candy machine), worked in the afternoon, ran errands, and now you're late for a meeting of your study group. Of course, you're starving! On the way to the library, you stop off at your favorite fast food restaurant to refuel. You order a double hamburger with cheese, a large order of fries, and a chocolate milk shake.

One person in your study group is taking a nutrition course and has been studying dietary analysis. He was shocked to discover that his normal intake wasn't even close to the guidelines for a healthy diet. As he tells you about this, you wonder whether your diet is any better. Below, you'll find the instructions provided by your friend for calculating the percentage of carbohydrates, protein, and fat in your diet. He also thoughtfully provided you with a copy of the government recommendations for a healthy diet (**Table 6-4**). **Table 6-5** summarizes your meals.

T A B L E 6 - 4	
RECOMMENDED DIETARY GUIDELINES	
NUTRIENT	RECOMMENDED PERCENTAGE OF DAILY CALORIC INTAKE
Carbohydrates	55–60%
Protein	10–15%
Fat	no more than 30%
1 g carbohydrate contains 4 kcal	1 g protein contains 4 kcal 1 g fat contains 9 kcal

Follow the instructions below to calculate the percentage of carbohydrate, protein, and fat you consumed today.

1. Begin your calculations with the hamburger. **Multiply** the number of grams of carbohydrates in the hamburger by **4** (since there are 4 kcal per gram of carbohydrate).

    Record the answer in Table 6-5. Perform the same calculations for the **protein** and **fat** in the hamburger (using the appropriate number of kilocalories) and record your answers.

2. Calculate the **kilocalories** for the other foods listed in **Table 6-5** and record your answers.

3. Total each column in **Table 6-5** and **record** your answers.

   Add the total **kilocalories** from **carbohydrates, protein,** and **fat** and record the answer at the bottom of **Table 6-5.**

TABLE 6-5 DIETARY CALCULATIONS						
Food Eaten	Carbohydrate (g)	Carbohydrate (kcal)	Protein (g)	Protein (kcal)	Fat (g)	Fat (kcal)
Double beef hamburger with cheese	54		51		60	
Fries (large order)	43		5		20	
Chocolate milk shake	49		9		10	
Bag of peanuts (2 ounces)	10		14		28	
Column Totals						
Total kcal:						

4. Calculate the percent of your diet that came from carbohydrates, protein, and fat.

$$\text{\% of carbohydrates in diet} = \frac{\text{total kcal from carbohydrates}}{\text{total kcal consumed}} \times 100 = \underline{\hspace{1.5cm}}\%$$

$$\text{\% of protein in diet} = \frac{\text{total kcal from protein}}{\text{total kcal consumed}} \times 100 = \underline{\hspace{1.5cm}}\%$$

$$\text{\% of fat in diet} = \frac{\text{total kcal from fat}}{\text{total kcal consumed}} \times 100 = \underline{\hspace{1.5cm}}\%$$

$$\overline{\hspace{1.5cm}100\hspace{0.3cm}}\%$$

5. Compare your calculated percentages to the federal government's recommended values.

   Was your intake today in line with recommended dietary goals?_____

   If not, which nutrients didn't match the guidelines?

6. Place an X in front of any of the following that are good suggestions for improving your day's food intake.

___ Bring an apple or an orange for a snack instead of the peanuts.

___ Replace the fries with a baked potato.

___ Replace the hamburger with pizza.

___ Replace the milk shake with low-fat milk or juice.

___ Eat a candy bar instead of the peanuts.

7. If your body requires **1500 kcal per day,** how many excess kilocalories did you eat in the day described in **Table 6-5?** _____ kcal

If you had consumed the same amount of excess kilocalories each day for the last three weeks, **how many excess kilocalories** have you accumulated? _____ kcal

Every time you accumulate 3500 excess kilocalories, you gain a pound.

How many pounds have you gained over the last three weeks? _____ lbs

8. You want to lose this excess weight, but with your schedule, you're too busy to exercise regularly. However, if you park in the furthest lot from the building, you can get to class in 15 minutes of quick walking. Every time you do this, you'll burn **150 kcal.**

If you walk quickly **to and from** the parking lot once a day, you can lose the weight you gained in only _____ days.

Check your answers with your instructor before you continue.

ACTIVITY 5   READING FOOD LABLES

You're sitting at breakfast with your bowl of cereal or toast and notice that nutritional information is included on the package label. This information is mandated by the **U.S. Food and Drug Administration (FDA).** The **food label** is a valuable tool for evaluating packaged food products. You can compare the information on a food label with the federal dietary guidelines you learned about in the previous activity to make diet selections that follow those recommended for good health.

Read the food label in **Figure 6-2** and answer the following questions.

This section lists the number of calories in one serving and also the number of fat calories per serving.

All the information on a nutritional label is per serving; however, many packages contain multiple servings.

Nutrition Facts

Serving Size 1 cup (228g)
Servings Per Container 4

Amount Per Serving

Calories 260   Calories from Fat 120

	% Daily Value*
Total Fat 13g	**20%**
Saturated Fat 5g	**25%**
Cholesterol 30mg	**10%**
Sodium 660mg	**28%**
Total Carbohydrate 31g	**10%**
Dietary Fiber 0g	**0%**
Sugars 5g	
Protein 5g	

Vitamin A 4%	•	Vitamin C 2%
Calcium 15%	•	Iron 4%

*Percent Daily Values are based on a 2,000 calorie diet. Your daily values may be higher or lower depending on your calorie needs:

		Calories:	2,000	2,500
Total Fat	Less than		65g	80g
Sat Fat	Less than		20g	25g
Cholesterol	Less than		300mg	300mg
Sodium	Less than		2,400mg	2,400mg
Total Carbohydrate			300g	375g
Dietary Fiber			25g	30g

Calories per gram:
Fat 9 • Carbohydrate 4 • Protein 4

Use the percent daily values to easily compare products and to quickly tell if a serving of food is high or low in nutrients.

The listed nutrients are those considered to be most important for good health.

INGREDIENTS: WATER, ENRICHED MACARONI (ENRICHED FLOUR, NIACIN, IRON, THIAMINE MONONITRATE, AND RIBOFLAVIN), EGG WHITE, FLOUR, CHEDDAR CHEESE (MILK, CHEESE CULTURE, SALT, ENZYME), SPICES, MARGARINE (PARTIALLY HYDROGENATED SOYBEAN OIL, WATER, SOY LECITHIN, MONO AND DIGLYCERIDES, BETA CAROTENE FOR COLOR, VITAMIN A PALMITATE), AND MALTODEXTRIN

Ingredients are listed in descending order of weight. The sources of some ingredients, such as certain flavorings, are stated by name to make it easier for people to identify ingredients that they wish to avoid for health, religious, or other reasons.

FIGURE 6-2. Sample Food Label

1. If you ate all the food in the labeled container, how much of each of the following would you have eaten? Enter your calculations in **Table 6-6** below.

TABLE 6-6 FOOD LABEL CALCULATIONS			
Total fat (g)		Total carbohydrate (g)	
Saturated fat (g)		Dietary fiber (g)	
Cholesterol (mg)		Sugar (g)	
Sodium (mg)		Protein (g)	

2. Calculate the fat calories recommended each day if you were on a **2000 calorie diet** (remember, the maximum should be 30% of your calories from fat).

    Recommended calories from fat:

    $2000 \times .30 \ =$

3. Calculate the percentage of "allowed" dietary calories from fat you would consume if you ate the entire container of macaroni and cheese discussed above.

    Percentage of calories from fat eaten:

    $$\frac{\text{fat calories eaten}}{\text{recommended calories from fat}} \ =$$

4. Would you consider this food a good choice if you were trying to control the amount of fat in your diet? **Explain** your answer.

5. If the number of calories from fat (listed in the upper right hand corner of the food label) was missing, how would you determine the approximate number of fat calories per serving?

Self Test

1. You add **Biuret solution** to your morning orange juice. The Biuret solution does not change color. What can you **conclude** from this experiment?

2. What do you think would happen if you placed a drop of iodine on your baked potato at dinner?

    On your steak?

    Explain your answer.

3. Match the foods in **Table 6-6** with their functions in the living organism they come from. For example, why does a cow produce milk? (**Hint:** Not for your morning cereal!)

TABLE 6-6	
MATCHING FOODS AND THEIR FUNCTIONS	
FOOD	FUNCTION IN PLANT OR ANIMAL
Lettuce	
Hamburger	
Tuna	
Milk	
Refried beans	
Peanut butter	

Use the food lable below to answer the following:

Nutrition Facts

Serving Size 1 oz. (28g)
Servings Per Container about 4

Amount Per Serving

Calories 125 Calories from Fat 40.5

	% **Daily Value***
Total Fat 4.5g	**7**%
Saturated Fat 0g	**0**%
Trans Fat 0g	**0**%
Cholesterol 0mg	**0**%
Sodium 70mg	**3**%
Total Carbohydrate 18g	**18**%
Dietary Fiber 2.25g	**8**%
Sugars 0g	
Protein 2g	

Vitamin A 5%	•	Vitamin C 25%
Calcium 0%	•	Iron 4%

4. How many milligrams (mg) of sodium are in one serving of this product?
_____

5. How many calories would you consume if you ate the entire contents of this package? _____

6. How many grams of fat are in one servings of this product? _____

7. How many calories from protein are in three servings of this product?
_____

8. How many servings of this product would you need to consume in order to obtain the minimum daily requirement of Vitamin C? _____

The Skin: Example of an Organ

Objectives

After completing this exercise, you should be able to:

- list and explain several features present in the epidermis that provide for protection of the body
- explain how each skin layer is different from the others
- identify each structure in the dermis and explain its function
- list and explain at least four examples of how deposits of fats and oils are useful to plants, animals, and microorganisms
- discuss several adaptations of epithelial tissues that demonstrate the relationship between structure and function
- discuss the relationship between the density of touch receptors in various body locations and touch sensitivity
- apply your knowledge of the special features of the skin to practical situations

CONTENT FOCUS

Within our bodies, we have approximately 100 trillion cells. As you observed in previous exercises, not all cells have the same **structure.** In the bodies of multicellular animals such as humans, not all cells have the same **function.**

A **tissue** is a **group of similar cells that perform a specific function.** Cells are organized into **four major tissue types:**

Epithelial Tissue	Lines all inner and outer body surfaces; covers organs and body cavities
Muscle Tissue	Contracts to produce movements
Connective Tissue	Joins and supports other tissues
Nervous Tissue	Senses stimuli; transmits signals around the body

Tissues group together to form the organs of the body. All of these tissue types are present in the **skin,** the body's largest organ.

To examine tissue function, you'll take a closer look at the **design of your arm,** beginning with the outer epidermis and working your way through the underlying tissues to the supporting skeleton.

ACTIVITY 1   SKIN: THE OUTER PROTECTIVE LAYER

1.   Work in **groups.** Set up a **dissecting microscope.**

2.   Place your hand, **palm-side down,** on the stage of the microscope. **Examine your skin** closely, beginning with the **lowest** magnification and gradually zooming up to the **highest.**

     Record your skin **observations.**

     I see the lines of my skin and hair in more detail

3.   **Remove your hand from the microscope stage.** Get a **dropper bottle of water.**

     While holding your hand **level** in front of you, **palm-side down (NOT under the microscope),** place a drop of water on the back of your hand.

     What happens to the water droplet?

     It stay in a drop on the surface

     Gently **tilt** your hand. **What happens** to the water droplet now?

     It moves across the surface of my skin

     Did the liquid penetrate the skin easily? *NO*

     Is this one of the **protective functions** of the skin? *Yes* **Explain** your answer.

     dis courage foreign invaders

4. The **waterproofing** quality of the skin is due to the presence of the **protein keratin** in epithelial cells. The **keratinized** cells of the epidermis also protect the underlying tissues from **mechanical injuries and abrasions.** On parts of the body where **abrasion is most common,** the epidermal layer tends to be **thicker.**

5. The skin also provides protection from exposure to **ultraviolet (UV) radiation.** Pigment-producing cells in the epidermis manufacture the **protein melanin** that blocks penetration of UV rays and protects underlying cells from damage.

    Most people have about the same number of melanin-producing cells, called **melanocytes. Dark-skinned** individuals, however, produce **more and darker melanin** than fair-skinned individuals do. Melanin production is also **stimulated by exposure to sunlight.**

    Explain how the following **factors work together** to form a protective barrier in the skin: **ultraviolet radiation, melanocytes, melanin, and a suntan.**

    all 4 combined together can increase risk
    of skin diseases such as cancer

    People with darker skin have a lesser chance
    of getting these because they have melanin
    in their skin

6. Problems with melanin production can result in different disorders.

    Albinos have a normal number of melanin-producing cells, but the cells don't synthesize melanin. For this reason, albinos have no pigment in their skin, eyes, and hair.

    In the condition **vitiligo,** melanocytes die, causing patches of skin to lose their coloration. These light spots in the skin are often surrounded by skin with normal pigmentation.

✓ Comprehension Check

1. Which of the following areas of the body probably have a thick epidermal layer? **(Circle ALL correct answers.)**

    a. inside of the cheeks   d. scalp
    b. palms of the hands   e. elbows
    c. soles of the feet   f. abdomen

2. The thick layers of keratinized cells mentioned in number 4 above form highly distinctive patterns of **ridges and whorls** that can be used to identify a person.

    We call these patterns _____.

3. To raise some money for your tuition, you take a summer job working construction. After several weeks, you notice that **calluses** have developed on your hands. What is a **callus? Why did the calluses form?**

4. The incidence of **skin cancer** is increasing in our population. **(Circle one answer.)** You have better protection from exposure to the damaging effects of UV rays if you have **dark / fair** skin.

Check your answers with your instructor before you continue.

ACTIVITY 2   THE EPIDERMIS

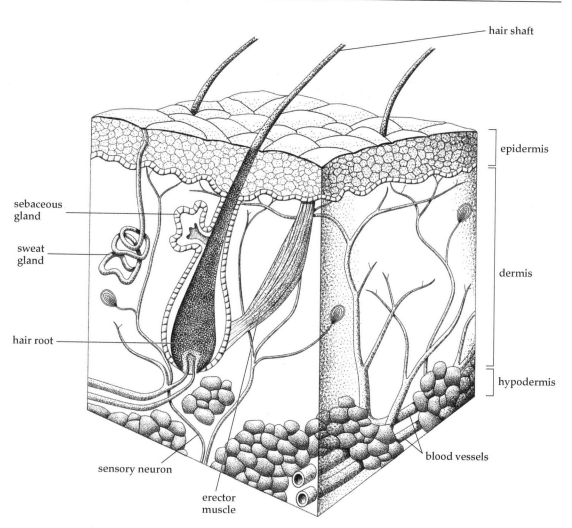

FIGURE 7-1.   Layers of the Skin

By looking at **Figure 7-1,** you can see that the skin is made up of several layers: the **epidermis** (the outermost layer), the **dermis** (the middle layer), and the **hypodermis** (the innermost layer).

1. Look at **Figure 7-2** and note that the **epidermis** consists of many layers of **epithelial cells.**

   Despite the fact that the epidermis looks very thick in **Figure 7-2,** you can see in **Figure 7-1** that the epidermis comprises only a small portion of the total thickness of the skin. The protective layer of keratinized cells at the **surface** of the epidermis is **repaired by rapidly dividing cells** at the **base** of the epidermis. Newly formed cells are gradually pushed upward to replace lost and damaged cells in the outer layer. In some individuals, excessive shedding of cells occurs in the scalp area.

   These discarded scalp cells are referred to as ___dandriff___ .

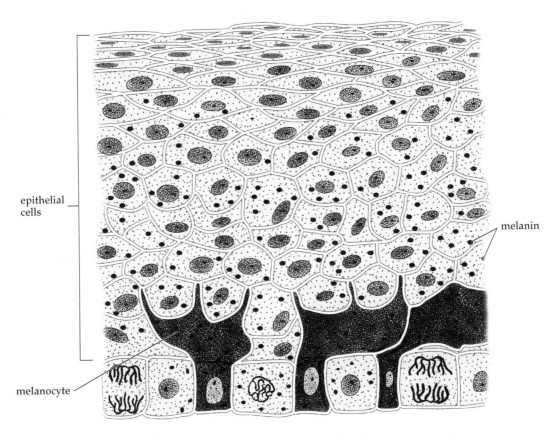

FIGURE 7-2. The Epidermis

☑ Comprehension Check

1. On **Figure 7-2, draw an arrow** showing the **direction of cell replacement** in the epidermis.

2. **Typical skin cells** are quite similar to those you've seen before. The nucleus appears as a shaded circle within the cell.

   In comparison, **dividing cells** have **distinctly visible chromosomes** within them (these appear as dark lines).

   On **Figure 7-2**:

   Draw an arrow that points to the **nucleus of a** cell that's **not** dividing.

   Draw an arrow that points to visible chromosomes within a **dividing** cell.

   Using a **colored pencil or highlighter, color in two cells** that are dividing.

3. **(Circle one answer.)** Actively dividing cells can be found in the **top / middle / bottom** layer of the epidermis.

4. On **Figure 7-2**:

   Place an **"O"** inside a cell that would be one of the oldest cells in the epidermis.

   Place a **"Y"** inside a cell that is one of the youngest cells in the epidermis.

   Place a **"B"** inside a cell close to a blood vessel.

5. The numerous **melanin molecules** shown in **Figure 7-2** were produced by cells called _____.

6. If the prefix **"epi"** means *above,* where do you predict the **dermis** of the skin will be located?

Check your answers with your instructor before you continue.

ACTIVITY 3   THE DERMIS

1. By looking at **Figure 7-1, circle the correct answer** in each of the following statements.

   The dermis is located **above / below** the epidermis.

   The dermis is **thicker / thinner** than the epidermis.

   Hair roots are located in the **dermal / epidermal** layer of the skin.

2. Take a look at the epidermal and dermal layers in **Figure 7-1.**

   Are any **blood vessels** present in the **epidermis?** _____

   If you cut **just** your epidermis, will the cut bleed? _____ **Explain** your answer.

3. Blood vessels carry nutrients and oxygen to body cells.

   (Circle one answer.)

   Epidermal cells at the **base** of the epidermis have **good / fair / poor** access to oxygen and nutrients.

   Epidermal cells at the **surface** of the epidermis have **good / fair / poor** access to oxygen and nutrients.

   The farther the epithelial cells are removed from good access to nutrients and oxygen, the **better / worse** their chances of survival.

4. **Blood vessels** near the root area supply the hair with **oxygen and nutrients** needed for growth.

   Melanin granules in the center of the hair shaft give the hair its color.

   Notice that **sebaceous glands** are located along the hair shaft. Oily secretions from the sebaceous glands **lubricate** the hair and skin, help prevent the skin from **drying out,** and **inhibit growth of bacteria.**

✓ Comprehension Check

1. The sale of **hand lotion** is a multimillion-dollar market in the United States. Using your knowledge of how the epidermis is lubricated, explain why hand lotion is such a big-selling item.

2. Explain why the skin of a person's scalp is often more oily than the skin of his/her arm.

 Check your answers with your instructor before you continue.

ACTIVITY 4   HAIR TODAY, GONE TOMORROW

Millions of hairs are scattered over the body. The average scalp has about 100,000 hairs and a man's beard has another 30,000.

1. Refer back to **Figure 7-1.** You'll see that a hair consists of two parts, the **shaft** and the **root.** The shaft portion of the hair begins in the **dermis** and extends **outside** the surface of the skin.

   The **root** is found **only within** the dermis and is surrounded by several layers of cells. Collectively, this structure is called a **hair follicle.**

2. Sit with your **eyes closed** while your partner runs his/her hand over the top of your head, **gently** touching the **hairs.**

   Can you tell when you are being touched? _____

   What **section** of the hair was just touched? _____

   A knot of **sensory nerve endings** is wrapped around the **root of each hair.**

   Moving a hair activates these nerve endings, so that your hairs function as sensitive touch receptors.

3. The outer part of the shaft contains heavily **keratinized** cells, which appear roughened and scale-like under a microscope.

   Hair pigment is made by **melanocytes** at the base of the hair follicle. Several types of pigments combine to produce the various hair colors.

   (Circle one answer.) Brown / red / black / blond hair has the most melanin present.

4. Living cells **divide** at the base of the follicle, causing your hair to grow. **Blood vessels** provide follicle cells with the oxygen and nutrition needed for growth.

5. Refer back to **Figure 7-1.** You can see that each hair is connected to a **tiny erector muscle** that **automatically** raises the hair when you're cold.

   In animals with fur coats, raised body hairs **hold a layer of warm air** next to the skin, insulating them from the cold. Humans have too few body hairs for this response to help keep us warm, but we are like other animals in our response to changing temperatures.

✓ Comprehension Check

1. Which of the following skin features can be found in the **dermis? (Circle ALL correct answers.)**

   a. melanin
   b. blood vessels
   c. nerve endings
   d. sebaceous glands
   e. erector muscles

   f. hair follicle
   g. hair shaft
   h. highly keratinized cells
   i. sweat glands

2. If you get a paper cut that penetrates to the **dermis,** will the cut bleed? _____ **Explain** your answer.

3. If the paper cut **only** penetrates the **epidermis,** will the cut **hurt?** _____ **Explain** your answer.

4. Your new pair of shoes is rubbing against your heel. If you walk around in them for several hours, it may cause an injury in which the epidermis separates from the dermis.

    When the space between the epidermis and the dermis fills up with fluid, it's called a _____.

Check your answers with your instructor before you continue.

ACTIVITY 5   BELOW THE SKIN (THE HYPODERMIS)

1. Beneath the skin is a layer of **connective tissue** with many fat cells. (See **Figure 7-1.**) This layer is called the **hypodermis** (**"hypo"** means *under* the dermis) or **subcutaneous** layer (**"sub"** means *under* and **"cutaneous"** refers to the skin).

    Perform the following experiment using a **"blubber mitten"** to discover one of the functions of the subcutaneous fat layer.

2. Work **individually.** Get the following supplies: **one blubber mitten and one empty mitten.**

    The two mittens are designed to simulate how the subcutaneous layer of the skin would function **with and without fat cells.**

    In this experiment, you'll place **one mitten on each hand** and **immerse both hands** in an **ice-water bath.**

3. **Before proceeding,** form a hypothesis about the **effect of water temperature** on each of your hands. **Record** your hypothesis below.

HYPOTHESIS FOR BLUBBER MITTEN EXPERIMENT

4. Go to the ice-water bath and **perform your experiment. Record your results** below.

RESULTS OF BLUBBER MITTEN EXPERIMENT	
WITHOUT FAT	WITH FAT

5. From the results of your experiment, what **conclusions** can you draw about the effect of fatty tissue in the subcutaneous layer of the skin?

✔ Comprehension Check

1. Which of the two mittens was the **control** in your experiment? _____

2. Why was it **necessary** to use a **control mitten** (instead of putting your **bare hand** in the ice-bath)?

3. **Based on the results** of your experiment, give **one function** of subcutaneous fat tissue.

4. Again, **referring to the results** of your experiment, suggest **two reasons** why some animals eat a lot in the fall season?

   a.

   b.

5. You're probably familiar with the fact that, when you **mix oil and water,** the oil will float to the surface. Do you think that significant **fat** deposits, such as those found in marine mammals, contribute to their **buoyancy? Explain** your answer.

6. Often **injections** of medicine are administered into the **subcutaneous layer** of the skin.

   The familiar term _____ **needle** refers to the location in the skin where the needle is inserted.

   Check your answers with your instructor before you continue.

ACTIVITY 6   THE SENSORY FUNCTION OF THE SKIN

The skin acts as an interface between the body and the outside environment. Information about several aspects of the outside world is acquired through activation of sensory receptors in the dermis and epidermis of the skin.

Sensory neurons process and transmit incoming information to the **central nervous system (the brain and spinal cord).** Neurons vary in function and appearance, but all share a similar three-part structure: the **cell body, the axon, and the dendrites** (see **Figure 7-3**). **The arrows on the Figure 7-3** indicate the **direction** of impulse transmission (from the dendrites to the cell body, and then to the axon).

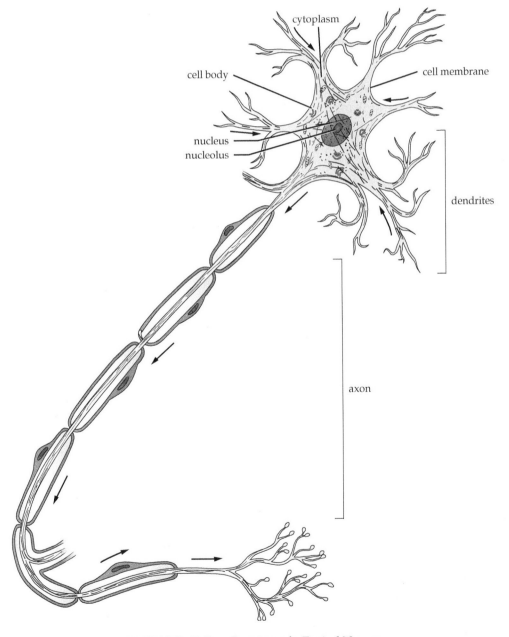

FIGURE 7-3. Structure of a Typical Neuron

Every neuron has a **cell body,** the region of the cell that contains the nucleus and a variety of other organelles necessary for cell metabolism. Outside stimuli are received by **dendrites** in the skin and other sense organs. Incoming signals from these dendrites are transmitted toward the cell body and continue into the **axon.** Axons relay outgoing messages from one neuron to other neurons or to various tissues and organs in the body.

Dendrites in the skin are specialized to receive different types of sensory input, such as **touch, temperature, pressure, pain, vibration, and proprioception** (the sense that tells you the current position of your body and limbs). In the following experiment, you'll be making observations about the distribution of touch receptors in the skin.

1. Work **with a partner.** Get the following supplies: **one set of touch calipers.**

    The set of touch calipers consists of a series of corks. Each cork has two pins inserted so that the blunt ends are protruding. The pins are spaced at varying distances from each other **(1 mm, 2 mm, 4 mm, 8 mm, 16 mm, and 32 mm apart).**

2. Form a hypothesis about the relative sensitivity of various body locations. For the **seven body locations** mentioned in **Table 7-1,** list them in order from the area that you think will be **least sensitive** to the area you think will be **most sensitive.**

TOUCH SENSITIVITY HYPOTHESIS

3. Perform the following set of tests on your partner, and then reverse roles and have your partner use the same tests to check your touch response sensitivity.

    Your partner should be seated, with his/her **eyes closed.**

    For each touch, the subject will report whether he/she feels one or two pins touching the skin.

4. Begin with the **32-mm caliper.** Lightly touch the skin on the back of the test subject's neck.

    Repeat the process with the remaining calipers in the set **(16 mm, 8 mm, 4 mm, 2 mm, and 1 mm) in any order you wish.**

For each test, record whether your partner is able to feel two pins touching the skin.

If the test subject felt **two pins**, record **yes**. If the test subject felt **one pin only**, record **no**.

Record these results in **Table 7-1**.

5. Repeat the same procedures at each of the following locations:

    back of the hand   tip of the nose

    palm of the hand   cheek

    inside of the forearm   tip of the forefinger

6. **Reverse roles** and have your partner repeat the same tests to check your touch response sensitivity. Record the results in **Table 7-1**.

T A B L E 7 - 1						
TOUCH SENSITIVITY IN VARIOUS BODY LOCATIONS						
TEST SUBJECT 1						
BODY LOCATION	32 mm	16 mm	8 mm	4 mm	2 mm	1 mm
Back of neck						
Back of hand						
Palm of hand						
Inside of forearm						
Tip of nose						
Cheek						
Tip of forefinger						
TEST SUBJECT 2						
Back of neck						
Back of hand						
Palm of hand						
Inside of forearm						
Tip of nose						
Cheek						
Tip of forefinger						

The **greater the number of touch receptors** in a body location, the **more sensitive** that location will be to touch stimuli. The reason for the increased sensitivity lies in the distribution of the receptors. If the pin points touch **two receptors that are adjacent** to each other, the pins will **activate both receptors together** and it'll feel as though you have been touched by **only one pin.**

If the pins touch two receptors that are **not adjacent,** however, each will **fire separately** and you'll feel the touch of **both pins.** The ability to feel the two pins is called **"two-point discrimination."**

In locations with a high density of touch receptors, therefore, the skin is much more sensitive to touch and much better at providing information about the number of touches (one pin or two pins). This type of touch sensitivity allows a surgeon or mechanic to perform delicate procedures working by touch alone.

The smallest pin distance when the test subject had effective two-point discrimination indicates the body location with the highest density of touch receptors.

The higher the number of touch receptors in a specific location, the greater the probability the pins can stimulate two nonadjacent touch receptors.

✓ Comprehension Check

1. Based on the results in **Table 7-1,** rank all areas tested in terms of their relative sensitivities to touch. List the body locations from the area of **least sensitivity** to the area of **highest sensitivity.**

    Subject 1:

    Subject 2:

2. Which of the body locations tested has the **greatest density** of touch receptors? **Explain** your answer.

3. Were there differences in touch sensitivity between you and your partner? If so, what differences did you observe?

4. Would you expect a decrease in touch sensitivity for each of the following conditions? **Explain each answer.**

   a. calluses on the palms of your hands:

   b. scar tissue:

   c. thick subcutaneous fat layer:

5. People who are visually impaired are able to read books and other printed materials by using the Braille system of writing. In the Braille system, words are represented by patterns of raised dots. Braille readers touch the dot patterns with their fingertips to read the text.

    How does the density of touch receptors in the fingertips make the Braille system possible?

Check your answers with your instructor before you continue.

Self Test

Fill in the blank with the choice that is **most appropriate** to describe the function of each skin structure. Answers can be used **only once.**

a. keratin
b. melanin
c. erector muscle
d. epithelial cell
e. epidermis

f. hair follicle
g. blood vessels
h. subcutaneous layer
i. sebaceous gland
j. dermis

1. _____ Produces oil to lubricate the hair and skin.

2. _____ Layer of skin containing many blood vessels and nerves.

3. _____ Type of protein deposits in the skin that form fingerprints.

4. _____ Type of dead cell that flakes off as dandruff.

5. _____ When the hairs on the back of a dog stand up, this tissue is responsible.

6. _____ Location of fatty tissue that insulates and protects the body.

7. _____ Accumulation of this protein helps protect your skin from the sun's rays.

8. _____ The outermost layer of the skin.

9. We often hear claims made by cosmetic companies that their **lotions** will **"make your skin young again."** Based on the information you've gained in this exercise, do you think that this is an accurate statement? **Explain** your answer.

10. You're determined to quit smoking, and so have just started using a nicotine patch on your arm. In which layer of skin does the nicotine first enter the bloodstream? **Explain** your answer.

11. In reference to the nicotine patch, even though it wouldn't be as visible, why would it be a bad idea to place the patch on the sole of your foot? (Answer in terms of the ability of the patch to **function**.)

12. **Challenge Question!** Burns are classified according to how many skin layers are destroyed. A **first-degree burn** affects only the outer layers of the epidermis. A **second-degree burn** destroys the epidermis and penetrates the dermis. **Third-degree burns** destroy skin tissues down to and including the subcutaneous layer.

    A nurse who works in the burn ward of a local hospital notices that patients who have second-degree burns suffer more pain than patients who have third-degree burns.

    Using your knowledge of the skin, **explain** why the more severe burn causes less pain.

The Musculoskeletal System

Objectives

After completing this exercise, you should be able to:

- identify three different types of muscle tissue using the microscope
- discuss the functions and body location of each type of muscle tissue
- explain the effects of muscle fatigue on muscle action
- identify the major structures of a long bone and explain the function of each
- discuss the contribution of minerals and protein fibers to bone strength and flexibility
- apply your knowledge of bone and joint structure to practical examples of skeletal support and movement

CONTENT FOCUS

In the previous exercise, you looked at the protective outer covering of the body. Now you'll move beneath the skin to examine muscles and bones.

Consider an Olympic sprinter at the beginning of a race. The moment the starting gun fires, muscles contract, the skeleton responds, joints allow for changes in body position, and tissues all over the body work together to coordinate forward movement. Tasks as simple as sitting, standing, walking, or taking notes in class all require similar coordination of muscles, bones, and joints.

ACTIVITY 1   BETWEEN THE SKIN AND BONES

1. If you were to take a **small piece of steak** and make a wet mount of the **muscle tissue,** you would see alternating light and dark bands. These bands, or **striations,** are composed of two proteins **(actin and myosin)** that are involved in **muscle contraction.**

2. To observe bands of actin and myosin, get a **prepared slide of skeletal muscle** (also called **"striated,"** meaning striped, muscle).

3. Observe the slide with the **compound microscope** on **high power.** Refer to the **photograph** in **Figure 8-1** to locate the following structures on your slide:

   a. **one** muscle cell (also called a muscle **fiber**)

   b. striations

   c. nuclei of **one** muscle cell

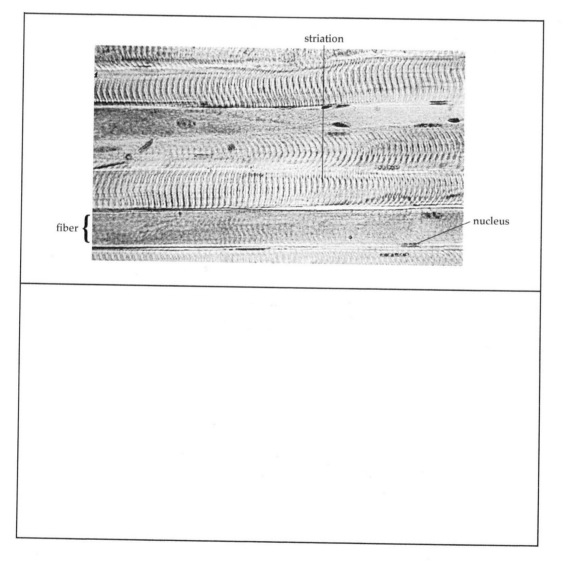

FIGURE 8-1. Skeletal Muscle Tissue

4. In the empty box in **Figure 8-1**, make a drawing of several skeletal muscle cells **as they appear in your microscope.**

5. **Label** the **striations, nuclei, cytoplasm, and cell membrane** of **one** skeletal muscle cell on your **drawing.**

 Check your answers with your instructor before you continue.

6. The **function** of skeletal muscle is to bring about **voluntary** movement. Skeletal muscles are under your conscious control. **Wiggle the fingers** of your left hand. You're using skeletal muscles!

7. There are **two other types of muscle tissue** that have **different functions** in the body. **Smooth muscle** allows for **involuntary** movements. Involuntary means that you have little control over its function.

    Observe the **two drawings** of smooth muscle in **Figure 8-2.**

    The protein fibers of smooth muscle cells are **not** arranged in bands, so **striations are not visible** in this tissue. Smooth muscles are found in the **digestive tract, blood vessels,** various **glands,** and many other locations where **automatic responses** are required.

8. **Add labels** for the following structures to the drawing of **individual smooth muscle cells** in **Figure 8-2: cell membrane, nucleus, and cytoplasm.**

9. **Cardiac muscle** is found **only** in the **heart.** It's responsible for your heartbeat. Refer to the photograph and drawing in **Figure 8-3.**

    Are cardiac muscle cells **striated** (similar to skeletal muscle cells)? _____

10. Contraction of all cardiac muscle fibers **must be coordinated** in order to maintain a regular heartbeat.

    To accomplish this coordination, cardiac muscle cells are tightly connected into a communication network that sends the signal to contract from cell to cell. These cell junctions are called **intercalated discs.**

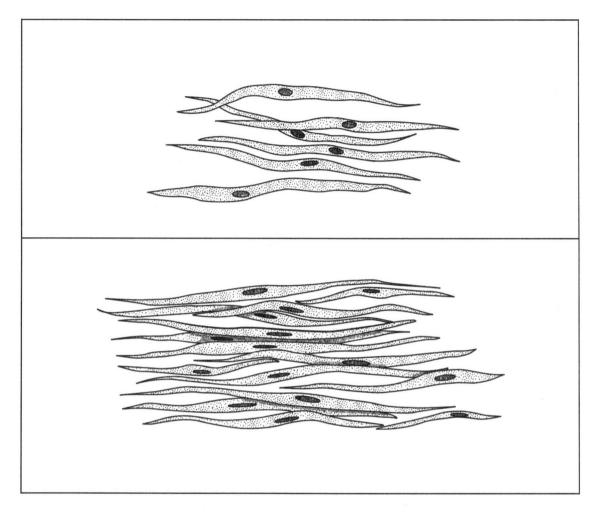

FIGURE 8-2. Smooth Muscle Tissue

✓ Comprehension Check

Now that you're familiar with the **three different types of muscle tissue,** answer the following:

Circle all answers that apply. Questions may have more than one correct answer.

1. The erector muscles that raise hairs on your arm consist of **smooth / skeletal / cardiac** muscle tissue.

2. When you **wink** your eye, you're using **smooth / skeletal / cardiac** muscle tissue.

3. When you're overheated, extra blood is directed to the skin for cooling. This occurs through the action of **smooth / skeletal / cardiac** muscle tissue.

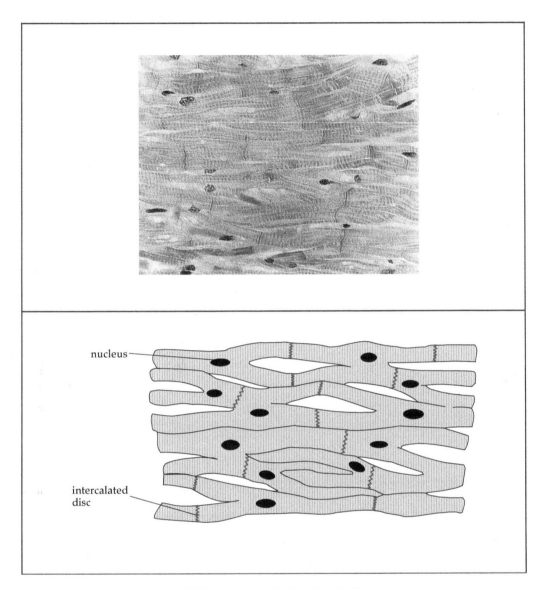

FIGURE 8-3. Cardiac Muscle Tissue

4. During a heart attack, **smooth / skeletal / cardiac** muscle tissue may be damaged.

5. Labor and childbirth involves the action of **smooth / skeletal / cardiac** muscle tissue. **Explain** your answer.

6. List **two** locations of **smooth muscle** in the body that were **not** mentioned in this exercise.

Check your answers with your instructor before you continue.

ACTIVITY 2   MUSCLE FATIGUE

During exercise, contracting muscles become progressively weaker until the muscle cells no longer respond to stimulation. This process is known as **muscle fatigue.** Muscle fatigue can be explained by several factors that occur simultaneously as your body works. These include:

- lack of ATP to meet energy needs
- insufficient oxygen for cell respiration
- depletion of energy reserves in the muscle cells

1. Work in groups. During this experiment, you'll be supporting a **heavy** book on your open palm with your arm completely extended. Which arm can support the book longer? **Form a hypothesis** about the ability of the muscles in your right and left arms to support the book and **record** your hypothesis below.

HYPOTHESIS FOR MUSCLE FATIGUE EXPERIMENT
Domanite arm for me that would be the right arm

Check your hypothesis with your instructor before you continue.

2. For your muscle fatigue experiment, one member of the group will act as a **timekeeper,** the second will **record** the experimental results, and the third will be the **test subject.**

Note:
The person holding the book shouldn't know how much time has expired until the entire experiment is completed.

---

3. Get an **extremely heavy book.**

   a. Place the book on your open hand.

   b. With the book on your hand, **extend your arm fully with the palm up. Don't bend your elbow.**

      Your arm should **not** be braced against your body.

   c. Record the length of time you can hold your arm extended. **Record** the results in **seconds** under **Trial 1** in **Table 8-1.**

   d. Rest your arm for **5 seconds** and repeat the experiment for **Trial 2.**

   e. Rest your arm for **5 seconds** and repeat the experiment for **Trial 3.**

   f. Rest your arm for **5 seconds** and repeat the experiment for **Trial 4.**

TABLE 8-1		
RESULTS OF MUSCLE FATIGUE EXPERIMENT		
TRIAL NUMBER	TIME ARM HELD EXTENDED (sec) RIGHT ARM	TIME ARM HELD EXTENDED (sec) LEFT ARM
1	1.45 min	1.30 min
2	1.30 min	1.10 min
3	1.15 min	.54 sec
4	1.10	.50 sec

4. **Graph** your experimental results in **Figure 8-4.** Plot **time** on the **Y-axis.**

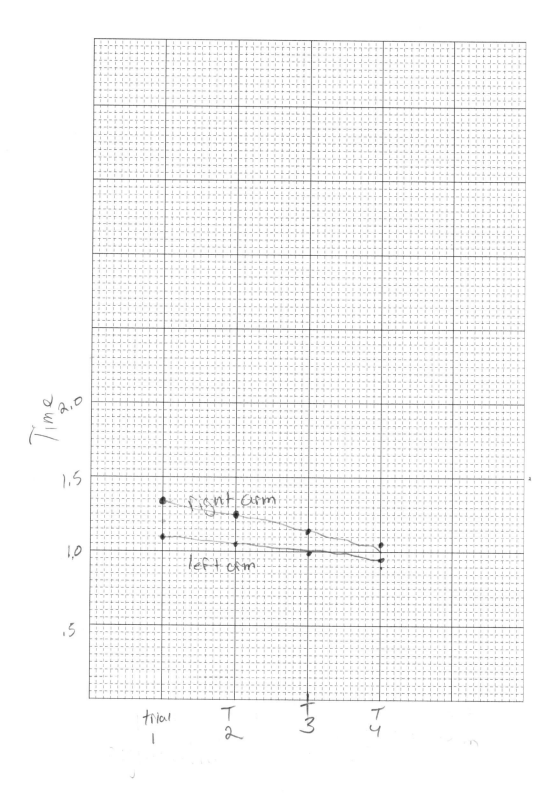

FIGURE 8-4. Comparison of Muscle Fatigue in Right and Left Arm

5. Did the experimental results **support** your hypothesis? _____Yes_____

   Explain your answer, mentioning **facts** collected during your experiment.

   The domante arm had a longer time holding the book

6. Did you see evidence of **muscle fatigue? Explain** your answer.

   arm started to shake
   You could see the book lower down

7. To have full confidence that your conclusions about muscle fatigue are accurate, **what changes would you recommend** for the experimental design?

   You could have a Yard stick virctially to see how much the arm will drop.

 Check your answers with your instructor before you continue.

ACTIVITY 3   SKELETAL SUPPORT

Bone is a type of **connective tissue.** Although bone may not resemble other body tissues superficially, it **is** composed of **living cells.**

   Bone cells produce a rigid material that forms the structure generally associated with bones. This mineral **"matrix"** in humans is composed of **calcium phosphate.**

Interwoven throughout the rigid mineral matrix are **protein fibers** that add **strength and flexibility.**

Compact bone is strong and densely packed. In compact bone, the cells are organized into vertical structures that resemble **drinking straws.** If you try to crush a straw by pushing on both ends, it'll be hard to do. In a similar manner, bones resist the force of gravity and support your body weight.

In contrast to compact bone, **spongy bone** consists of a loose arrangement of thin plates of bone similar to the appearance of a sponge. **Spongy bone is lighter** than compact bone, **reducing the weight of the skeleton** and making it easier for the muscles to move the bones.

Bone strength is a result of compact and spongy bone working together. **Compact bone** can only resist forces applied **along the length** of the bone, while **spongy bone** helps resist stresses **applied from many directions** (such as a kick to the side of your leg).

The inside of a bone contains **bone marrow.** The central cavity of the bone **shaft** contains **yellow marrow,** which functions as a fat-storage area. At the **ends of the long bones,** gaps in the spongy bone contain **red marrow,** which produces all types of **blood cells.**

Bone cells, as with cells of all tissues, require delivery of oxygen and nutrients and removal of waste materials. Although you can't see them without magnification, **channels throughout the bone** provide passageways for a network of blood vessels and nerves to serve these cells.

1. Work in groups. Get a **long bone** that has been cut open lengthwise.

    Set up a **dissecting microscope** for your observations.

2. **Examine** the bone with and without the microscope.

    Based on your **observations, label** the following structures in **Figure 8-5: compact bone, spongy bone, shaft, marrow cavity, the location of red marrow, and the location of yellow marrow.**

3. **(Circle one answer.)** Most blood cells are produced in the **shaft / ends** of the bone.

    Explain your answer.

    Gaps in the spongy bone contain red marrow which produces all types of blood cells.

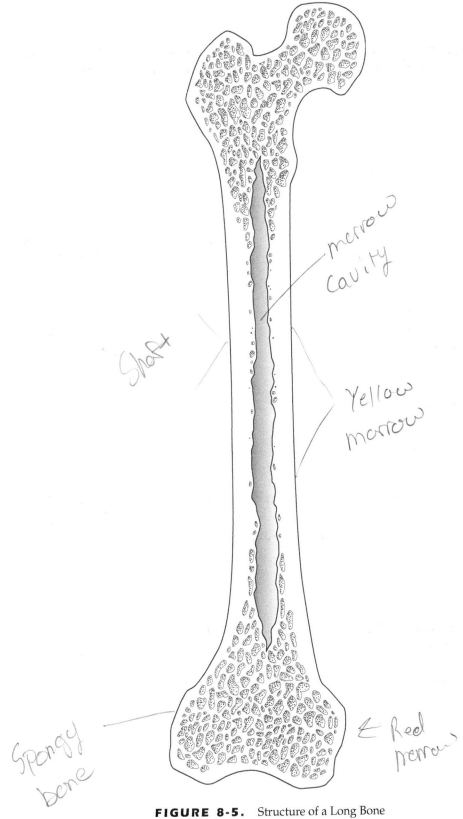

FIGURE 8-5. Structure of a Long Bone

4. **(Circle one answer.)** In situations of long-term starvation, **red marrow / yellow marrow** would probably decrease first. **Explain** your answer.

    Yellow because of fat cells.

5. Examine a **human pelvis,** which is composed of several attached bones.

    Based on the **function of the pelvis and its position in the skeleton,** form a **hypothesis** about the relative **amounts of compact and spongy bone** present in the bones of the pelvis. **Record** your hypothesis below.

HYPOTHESIS: COMPACT VS. SPONGY BONE IN THE PELVIS
Compact, more of a dense bone, that has to hold up to force.

6. **Explain** the line of reasoning you followed to develop your hypothesis.

    You have more force on your body which means you need a stronger support.

7. Suggest a method you could use to **test** your hypothesis.

    use a mri to see the bone.

8. **Challenge Question!** In many societies, heavy bundles are balanced on people's heads, rather than carried in their arms or on their backs (see **Figure 8-6**). **Explain** how **compact and spongy bone** each contribute to the ability of the skeleton to support the load.

Compact of the Skull is going to be more stable the the spongy bone.

Check your answers with your instructor before you continue.

FIGURE 8-6. Supporting a Load on Your Head

ACTIVITY 4
WEAK AND STRONG BONES: CLASSROOM DEMONSTRATION

Bone is about **75% minerals (calcium phosphate)** and **25% protein fibers (mostly collagen)**. These materials, **acting together,** give bone its strength and flexibility.

1. Your instructor has prepared several **chicken bones** that have been subjected to **three different treatments.** The bones will be used to demonstrate how **minerals and protein fibers** in bone tissue contribute to strength and flexibility.

    Dish A   Contains cleaned chicken bones that were soaked in acid for several days. Under these conditions, **minerals** are removed, but the **protein** content doesn't change.

    Dish B   Contains cleaned chicken bones that were baked in a very hot oven. Under these conditions, the **protein fibers** are destroyed, but the **mineral** content doesn't change.

    Dish C   Contains chicken bones that were cleaned and dried.

2. Which dish contains the **control** bones? ___C___

3. **(Circle one answer.)**

    The acid-treated bones are **more** / **less flexible** than the control bones.

    The acid-treated bones are **harder** / **softer** than the control bones.

    Explain your answer to both statements.

    The acid is going to break down the minerels leaving behind the protine.

4.   **(Circle one answer.)**

The baked bones are **easier / harder** to **break in half** than the control bones.

The baked bones are **easier / harder** to crush **into powder** than the control bones.

Explain your answer to both statements.

The bone Srinks in the oven when exposed to heat, and becomes easier to crush.

✓ Comprehension Check

1. In the condition known as **rickets,** the leg bones bend under the body's weight. Using your knowledge of the role of **minerals and protein fibers** in bone structure, explain one possible cause of rickets.

2. In the condition **osteoporosis,** the spongy bone of the pelvis and femur (long bone of the thigh) can be affected by a severe loss of bone mass. Using your knowledge of bone structure, explain why a person with this disease would be more susceptible to hip fractures.

Check your answers with your instructor before you continue.

ACTIVITY 5

A JOINT VENTURE: MOVING THE BONES OF THE SKELETON

Since bones are rigid, the skeleton can only change position through the action of joints. Most joints of the body are **synovial joints.** At synovial joints, the adjacent bones are separated by a fluid-filled capsule. Some joints, such as the knee, give us more problems than others. The problems originate because of weakness in the joint supports.

In the hip and shoulder joints, a **ball** on the end of one bone fits securely into a **socket** on another bone. The knee joint, in contrast, is held together only by tough bands called **ligaments.** The ligaments stretch like thick rubber bands between the ends of the thigh and shin bones, pulling them toward one another.

The **collateral ligaments** support the joint on the right and the left sides. The **cruciate ligaments** cross inside the knee from front to back, preventing the knee from overextending.

1.   Work in groups. Get a **knee model.**

2.   Locate the **two collateral and two cruciate ligaments.** Ligaments form attachments **between bones. Label** the **four support ligaments** on **Figure 8-7.**

3.   The knee joint connects the **femur** (thigh bone) and the main support bone of the lower leg, the **tibia** (shin bone). The smaller lower leg bone, the **fibula,** isn't part of the knee joint. Although the fibula doesn't help support body weight, it's an important surface for muscle attachment and provides stability to the ankle joint.

   Label the three leg bones on **Figure 8-7.**

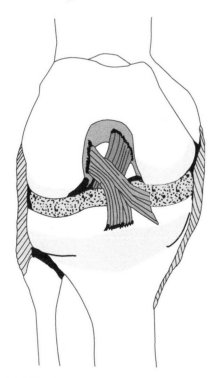

FIGURE 8-7.   Synovial Joint of the Knee

4. The **patella** (kneecap) protects the joint and prevents the knee from bending in the wrong direction.

5. The **meniscus**, a thick pad between the femur and tibia, **improves the fit** between the bone ends, making the joint more stable. Locate the **meniscus** on the knee model and label it on **Figure 8-7.**

6. Muscles and bones are connected by **tendons.** The ligaments and tendons of the knee form a **capsule** around the joint (see **Figure 8-8**).

7. No matter how smooth the ends of the bones may look, their surfaces are still uneven. When placed close together, these uneven surfaces rub against one another, producing heat, friction, and wear. In joints, the ends of the bones are covered with a **cushioning layer** of **articular cartilage.**

8. A membrane that lines the joint capsule, called the **synovial membrane**, produces **fluid** that lubricates the joint (called **synovial fluid**), reducing friction.

9. Although they aren't present on the model, the knee joint also has many fluid-filled sacs called **bursae** (singular **bursa**).

    These tiny bags of lubricating fluid reduce rubbing in areas where bones come in contact with ligaments, muscles, and skin.

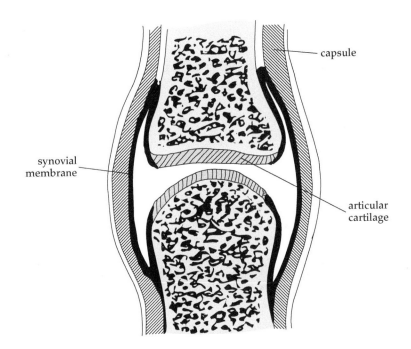

FIGURE 8-8. Interior of a Synovial Joint

✓ Comprehension Check

Fill in the blank with the **most appropriate** answer. Answers can be used **only once**.

a. meniscus
b. collateral ligament
c. articular cartilage
d. synovial membrane

e. synovial fluid
f. cruciate ligament
g. bursae
h. patella

1. _____ Covers the rough ends of bones, creating a smooth surface.

2. _____ A frequent football injury occurs when a player is hit on the side of the knee, tearing this structure.

3. _____ When this structure tears, there's nothing to prevent the knee from extending too far forward.

4. _____ If this structure, which provides a cup for the ends of the long bones, is damaged, the bones may slip out of position in the joint.

5. _____ Excess "fluid on the knee" is produced by this structure.

6. _____ Repetitive movement may lead to inflammation of these bags of cushioning fluid, producing swelling and pain.

Check your answers with your instructor before you continue.

Self Test

Fill in the blank with the choice that is **most appropriate** to describe the function of each in the body. Answers can be used **MORE THAN** once.

a. spongy bone
b. skeletal muscle
c. cardiac muscle
d. yellow marrow

e. red marrow
f. smooth muscle
g. compact bone
h. shaft

1. _____ A storage tissue for fat in bones.

2. _____ Tissue that produces blood cells.

3. _____ Densely packed tissue found at the outside of an arm bone.

4. _____ When the hairs on the back of a dog stand up, this tissue is responsible.

5. _____ When an emergency room doctor uses an electrical shock to start someone's heart, he/she is stimulating this type of muscle.

6. _____ When you touch a hot stove and jerk your hand away, this muscle type is contracting in your arm.

7. _____ This type of muscle moves the leg at the knee joint.

8. _____ Skeletons of birds adapted for flight would have more of this type of bone.

9. _____ This part of a long bone will be longer in taller people.

10. _____ If you placed an arm bone in a vise and applied pressure to both ends, this type of bone would resist breaking.

11. You and a friend are at the gym working out. Your friend is upset because she can't do as many repetitions of a bench press in her third set as she did in her first set. What would you tell her?

12. You're a medical examiner collecting bones from a building where there was a very intense fire. You find that many of the bones are brittle and crush easily. **Explain** this observation.

13. **(Circle one answer.)**

    The upper end of the patella is connected to a large muscle in the thigh by a **tendon / ligament.**

    The lower end of the patella is connected to the tibia by a **tendon / ligament.**

Examination of Skeletal Structure

Objectives

After completing this exercise, you should be able to:

- Identify the bones of the axial and appendicular skeletons when articulated and disarticulated
- Locate each bone using correct anatomical terminology
- Differentiate between left and right for paired bones from a disarticulated skeleton
- Determine the gender of an individual from the pelvic bones
- Use the technique of measuring femur length to estimate height
- Apply your knowledge of skeletal analysis to real-life situations

CONTENT FOCUS

The human body is composed of 206 bones divided into two groups: the **axial** skeleton and the **appendicular** skeleton.

- **Axial skeleton** — Bones that lie **near the** midline **(axis)** of the body.
- **Appendicular skeleton** — Bones of the **appendages** (such as arms, hips, shoulder, and legs).

These bones provide the framework for support of the human body. By studying the different bones of the skeletal system, scientists are able to gather information about individuals and populations. Patterns of disease, nutrition, migration, mortality, reproduction, and other factors can be explored by analyzing the remains of ancient cultures. Currently, there are several popular television programs that rely on forensic evidence to solve cases. The physical evidence they collect includes information gathered by the examination of bones.

Anthropologists and forensic scientists can use the information gathered from bones to help answer several questions, including:

- Which bones are present? Which are missing?
- Are these bones from one individual or from many?
- What was the sex of the individual?
- What was the approximate age of the individual at the time of death?
- What was the approximate height of the individual?

The answers to these questions can provide valuable leads to an investigating agency and help identify the remains of an unknown person.

ACTIVITY 1   THE HUMAN SKELETON

1. Work **in groups.** Using the information in **Table 9-1,** label all the bones identified on the diagram in **Figure 9-1.** Use the **correct terminology** (not the common names) for each bone you label.

2. Compare your labeled diagram with the articulated skeleton in the classroom. A picture may be worth a thousand words, but there's no substitute for the real thing.

TABLE 9-1

BONES OF THE ADULT SKELETON

AXIAL SKELETON	APPENDICULAR SKELETON
skull 22	**pectoral girdle (shoulder):**
hyoid 1	clavicle (collar bone) 2
auditory ossicles (inner ear) 6	scapula (shoulder blade) 2
vertebral column (spine) 26	**upper extremities:**
sternum (breastbone) 1	humerus (upper arm) 2
ribs 24	ulna (forearm) 2
	radius (forearm) 2
	carpals (wrist) 16
	metacarpals (body of hand) 10
	phalanges (fingers) 28
	pelvic girdle (hip):
	hip bones 2
	lower extremities:
	femur (upper leg) 2
	fibula (lower leg) 2
	tibia (shin bone, lower leg) 2
	patella (kneecap) 2
	tarsals (ankle) 14
	metatarsals (body of foot) 10
	phalanges (toes) 28
Total: 80	**Total: 26**

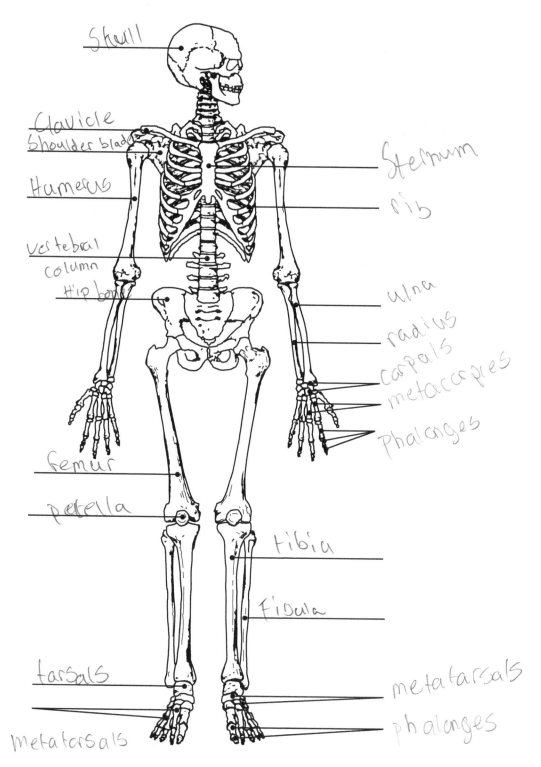

FIGURE 9-1. Bones of the Adult Skeleton

Check your Figure 9-1 labels with your instructor before you continue.

ACTIVITY 2   NAVIGATING THE BODY ANATOMY

Anatomical Position

In the standard anatomical reference position, the body is viewed from the front with the palms facing forward (see **Figure 9-2**). All descriptions of bones are based on this orientation of the human body.

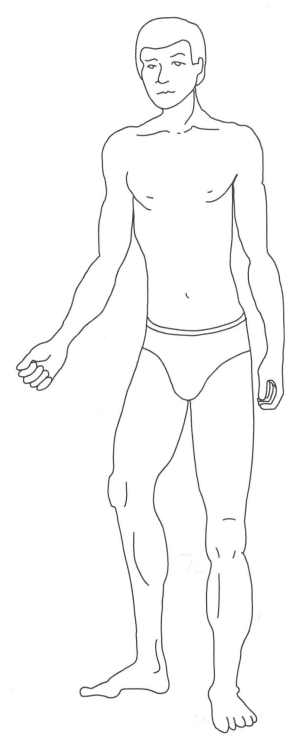

FIGURE 9-2. Anatomical Position

Definitions of Anatomical Terms

ANATOMICAL TERM	DEFINITION
TABLE 9-2 DEFINITIONS OF ANATOMICAL TERMS	
proximal	close to its attachment to the body (for example, the elbow is proximal to the wrist)
distal	away from its attachment to the body (for example, the fingers are distal to the wrist)
superior	above (for example, the shoulders are superior to the hips)
inferior	below (for example, the mouth is inferior to the nose)
medial	closer to the body midline (for example, in the lower arm, the ulna is medial to the radius)
lateral	away from the body midline (for example, the ear is lateral to the nose)
anterior	front side of the body (for example, the chest is anterior to the buttocks)
posterior	back side of the body (for example, the spine is posterior to the stomach)

> ### Note:
>
> In humans, the terms anterior and ventral both refer to the front side of the body (the side that faces forward when we walk). In four legged animals [such as the fetal pig], however, the front side of the body faces the ground, and thus is referred to as ventral, while the term anterior refers to the head end of the animal.

Using the information in **Table 9-2,** fill in the blanks with the correct anatomical terms:

1. The sternum is ___anterior___ to the heart.

2. The skull is ___superior___ to the hip bones.

3. The nose is ___medials___ to the ears.

4. The shoulder is ___proximal___ to the elbow.

5. The tarsals are ___distal___ to the tibia.

6.   The patella is _____ to the clavicle.

7.   The vertebral column is _____ to the sternum.

8.   The index finger is _____ to the ring finger.

Check your answers with your instructor before you continue.

ACTIVITY 3   TELLING YOUR LEFT FROM YOUR RIGHT

It's easy to differentiate between the left and right sides of the skeleton when it is artic-ulated. The problem, however, is that bones are often found **separated** from the rest of the skeleton. How can these **disarticulated (single)** bones be identified?

Each bone has certain characteristics that distinguish it from other similar bones. The appearance of various bone features and their locations on a particular bone are used by forensic anthropologists to determine if the bone came from the right or left side of the body.

1.   Work in groups. From the supply area, obtain a **numbered container of disarticu-lated bones.**

2.   Use the following information to **identify each bone** in your container.

Note:

Don't panic at the technical names for bone features. They are only to be used for identification and side determination. DO NOT memorize the names of all these features without consulting your instructor. Your instructor will help you determine which information is most important.

---

3.   **If possible, determine whether each of the following bones came from the right or left side of the body:** humerus, femur, and patella.

Side determination of a complete humerus (see Figure 9-3):

■ Hold the bone with the head up. Look at the opposite end of the bone. Notice that both sides have an indentation. The indentation that is deeper and larger in size is on the posterior surface. Place the bone over your own right or left humerus such that **posterior surface is touching your arm and the head is medial.** You have now identified the right or left humerus.

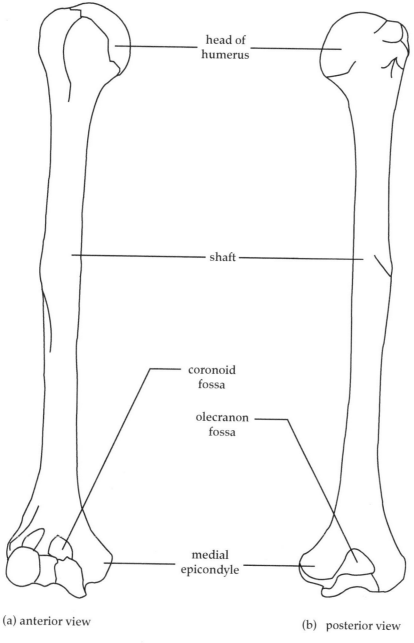

(a) anterior view   (b) posterior view

FIGURE 9-3.   Humerus with Identifying Characteristics Labeled

Side determination of a complete femur (see Figure 9-4):

▪ Hold the bone with the head up. Look at the opposite end of the bone. Notice that both sides have an indentation. The indentation that is deeper and larger in size is on the posterior surface. Place the bone over your own right or left femur such that **posterior surface is touching your thigh and the head is medial.** You have now identified the right or left femur.

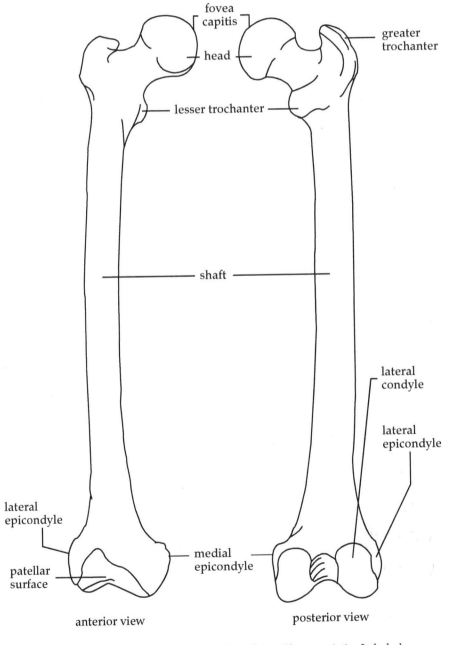

FIGURE 9-4. Femur with Identifying Characteristics Labeled

Side determination of a complete patella (see Figure 9-5):

■ Hold the bone with the posterior side facing you and the **apex** (pointed end) down. Note the two depressions (called **articular facets**) that mark the site of articulation with the base of the femur. **The larger (lateral) articular facet is on the side that the bone is from.**

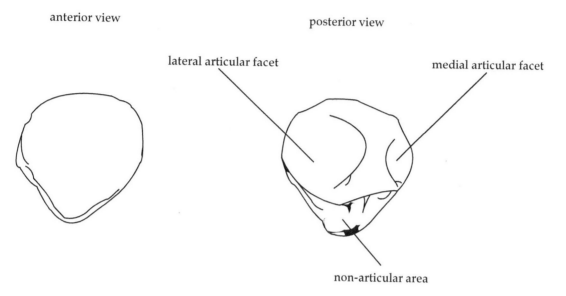

FIGURE 9-5. Left Patella with Identifying Characteristics Labeled

4. Record the results of your side determinations for the humerus, femur, and patella in **Table 9-3**.

TABLE 9-3
SIDE DETERMINATIONS OF DISARTICULATED HUMAN BONES

CONTAINER ID#:	
Name of Bone	**Side Determination (Left or Right)**
Femur	right
fibula	left
Patella	right
tibia	left
hip	right

Check your answers with your instructor before you continue.

✓ Comprehension Check

1. In a recent excavation of the city of Alexandria, scientists found a large collection of disarticulated human bones: however, there were no skulls. The inventory of the find is as follows:

individual vertebrae	62
clavicle	2
ribs	24
radius	3
sternum	1
femur	5
fibula	3
tibia	6
patella	1

Based on the inventory of bones discovered, what is the **minimum number of people** that could be represented in this collection? **Explain** your answer in detail.

Check your answers with your instructor before you continue.

ACTIVITY 4   DETERMINING GENDER FROM THE PELVIC BONES

You have probably noticed that women, generally, have broader hips than men. This characteristic is due to differences between the male and female pelvis.

The pelvis is composed of a pair of irregularly shaped bones called the **hip bones.** Each adult hip bone is formed by the fusion of three separate bones, **the ilium, the ischium, and the pubis,** which unite during adolescence (see **Figure 9-6**).

a. **Ilium — Upper** portion of the pelvis.
b. **Ischium — Lower** portion of the pelvis (supports the sitting body).
c. **Pubis — Anterior** portion of the pelvis. The anterior articulation of the hip bones is called the **pubic symphysis.**

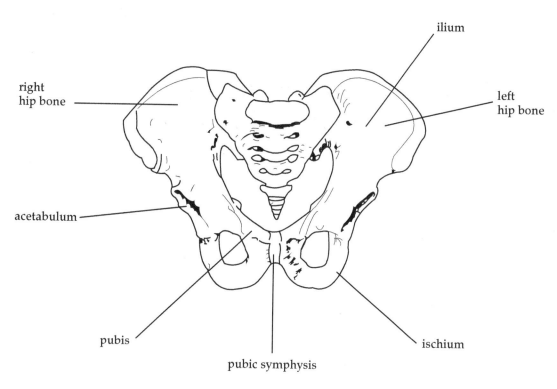

FIGURE 9-6. Bones of the Pelvis

Gender determination (see Figure 9-7):

1.   Work in groups. Select a pelvis from your box of disarticulated bones.

2.   In the **male hip bones,** the pubis is shorter, causing the subpubic angle to approximately resemble the **letter "V"** in an articulated pelvis.

-   Place your index finger flat against the pubic symphysis, allowing your thumb to rest in the subpubic space. If your thumb can move only slightly or not at all, the sex is probably **male.**

-   In the **female hip bones,** the pubis is longer, causing the subpubic angle to approximately resemble the **letter "U"** in an articulated pelvis.

-   If your thumb has more room for movement, this is an indication that the hip bones bone belongs to a **female.**

-   In general, the pelvis of a male is narrower, heavier and rougher than in females (for muscle attachment). The pelvic girdle of a female is wider, lighter, and smoother.

-   **The pelvic opening** is generally **large and oval in females** and **small and heart shaped in males.**

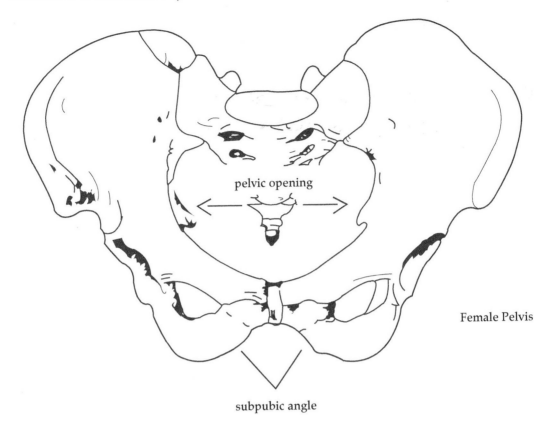

Female Pelvis

subpubic angle

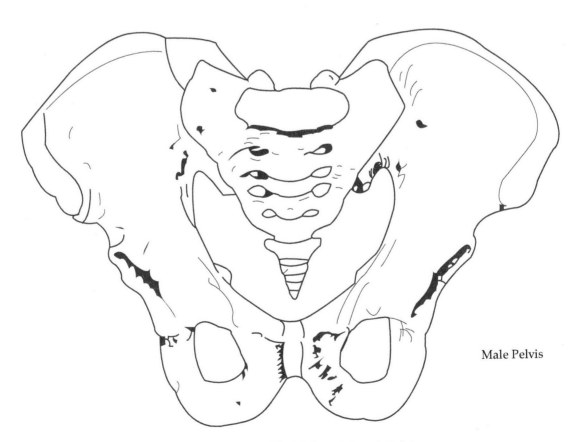

Male Pelvis

FIGURE 9-7. The Male and Female Pelvis

3. The **acetabulum,** the articulation site for the head of the femur, is **larger in males than females.** What does this tell you about the difference between the **head of the femur** in males and females?

*Femals would be wider/Skinners
males are open + More round.*

Gender determination guides using other bones of the body:

- Generally, male bones are larger, heavier, and rougher than female bones.
- The female skull is generally smaller and rounder than the male. It exhibits a vertically longer forehead and a smaller jaw.
- The female sacrum is broad and short compared to the longer, narrower male sacrum.
- The female coccyx (tailbone) is more moveable than the male coccyx.
- Several of the bones of the body can be measured and compared to reference standards for gender, height, and age estimation.

✓ Comprehension Check

1. You are an expert on skeletal remains. You receive a phone call from a local police department, which has recovered a pelvic bone that was found in the woods by a hiker. They would prefer not to waste time sending the bone through the mail, and ask you to describe to them, over the phone, the method they should use to determine if the bone came from a male or female. **Be clear and complete.**

2. Some bones were recovered from two burials in a cemetery with unmarked graves. Both skeletons had all the bones present. In the skeleton 1, the subpubic angle was clearly "V" shaped and the sacrum was long and narrow. In skeleton 2, the subpubic angle was clearly "U" shaped with a broad sacrum.

What was the gender of each individual?

If the pelvic bones were missing from the skeleton, could the number of ribs be used to determine the gender of each individual instead? **Explain** your answer.

Check your answers with your instructor before you continue.

ACTIVITY 5

ESTIMATING BODY HEIGHT FROM FEMUR LENGTH

An individual's height can be estimated from skeletal remains even if the skeleton is disarticulated. There are several different ways by which height can be estimated. In all cases, there is a margin of error (expressed as +/− in the calculated height of the individual), which must be taken into consideration when trying to identify the skeletal remains.

In this activity, you will be making height estimates based on femur length.

1.   Work in groups. Select a femur from your box of disarticulated bones. Get the following: **a tape measure.**

2.   Measure the diameter of the head of the femur, in **millimeters (mm).** Record your measurement: _____ **mm**

3.   Measure the mid-shaft circumference of the femur in **millimeters (mm).** Record your measurement: _____ **mm**

4.   Measure the length of the femur from the **greater trochanter to the lateral condyle in centimeters (cm).** Record your measurement: _____ **cm**

5.   Compare your femur measurements with the information in **Table 9-4** to determine the gender of the individual from which the femur came.

TABLE 9-4 DETERMINATION OF GENDER FROM FEMUR MEASUREMENTS			
MEASUREMENTS	FEMALE	GENDER DETERMINATION QUESTIONABLE	MALE
Diameter of head of femur	<43 mm	43–47	>47 mm
Mid-shaft circumference of femur	<81 mm	81 mm	>81 mm
Femur length			

Note:

If all the above femur measurements fall into the "Gender Determination Questionable" range, choose another femur and measure again.

6. What was the probable gender of the individual that this femur belonged to?
   _____

   Did both the femur and the pelvis in your box of disarticulated bones come from the same gender individual? **Explain** your answer.

7. **Based on your gender determination,** use the formula below and your femur length measurement to calculate the estimated height of the individual.

   for **male** skeletons:

   estimated height = (measured length of femur × 2.32) + 65.53

   for **female** skeletons:

   estimated height = (measured length of femur × 2.47) + 54.10

   > ### Note:
   > **For the male femurs, the margin of error in the femur length calculation is +/−6.96 cm. In females, the margin of error in the femur length calculation is +/−5.96 cm.**

8. Based on your femur calculations, how tall was the individuals the femur came from? _____ cm

   Converted to feet and inches _____

   (Hint: One inch = 2.54 centimeters.)

9. Explain how information on gender and height can be used to help identify a set of skeletal remains.

✓ Comprehension Check

1. A construction company is excavating a site for a new building. In the process of digging for the foundation, a few bones are discovered. Old stories say that a woman disappeared from a house that formerly stood on the site, and her ghost haunts the premises.

    Among the bones was a femur with a head diameter of 46 mm and a femur length of 50 cm. Have the construction workers found her remains? **Explain** your answer.

Check your answers with your instructor before you continue.

Self Test

Enter the letter of the correct bone for each of the following descriptions. Answers can be used **more than once** and questions may have **more than one** correct answer.

a. clavicle
b. carpals
c. femur
d. fibula
e. humerus
f. hyoid bone
g. hip bones
h. metacarpals
i. metatarsals
j. phalanges

k. patella
l. radius
m. ribs
n. scapula
o. sternum
p. tarsals
q. tibia
r. ulna
s. vertebrae

1. _____ Small bone that comes in contact with the ground when you kneel.

2. _____ Lateral bone of the leg.

3. _____ Bone that is medial and anterior to the ribs.

4. _____ Bones forming the knee joint.

5. _____ Bone affected if you bumped the anterior part of your lower leg.

6. _____ Shoulder blade.

7. _____ Collar bone.

8. _____ Breastbone.

9. _____ Medial bone of the leg.

10. _____ Bone that is distal to the clavicle.

11. _____ Bones of the ankle.

12. _____ Bones of fingers and toes.

13. _____ Medial bone of the forearm.

14. _____ Lateral bone of the forearm.

15. _____ Bones of the pelvis.

16. _____ Bones encompassed by your wristwatch.

17. _____ Bones inside your tennis shoe.

18. In an articulated skeleton, the ribs are connected to the sternum with cartilage. An artificial material replaces this cartilage in the classroom skeleton.

    Explain why this design functions better than having the bones of the rib cage connected directly to the sternum.

10

The Nervous System

Objectives

After completing this exercise, you should be able to:

- identify and explain the function of each of the major structures of the brain and spinal cord
- explain the effects of damage to various regions of the brain and spinal cord
- discuss the differences in function among sensory, motor, and interneurons
- diagram and explain the sequence of actions in a spinal reflex arc
- compare and contrast grey matter and white matter
- draw conclusions that are supported by experimental data
- analyze data using common statistical measures
- apply your knowledge of the scientific method to real-life situations

CONTENT FOCUS

To function properly, all of your organ systems must coordinate and communicate with the nervous system. Every activity in your daily life is made possible through the action of the nervous system. Since the nervous system plays such an integral role in your life, it is important to have a working knowledge of basic nervous system functions.

The **nervous system** responds **rapidly** to environmental stimuli. This rapid response time allows you to cross a busy street, react to emergency situations, read this book, or use a computer keyboard.

In this exercise, you will be examining the structure of the **central nervous system (the brain and spinal cord)**. You will also have a chance to explore various functions that are performed by these organs.

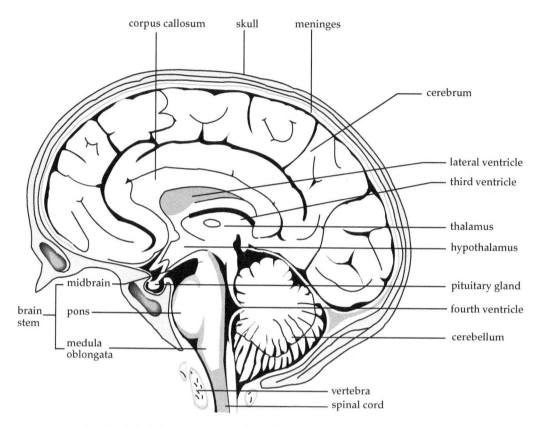

FIGURE 10-1. Longitudinal View of the Human Skull and Brain

ACTIVITY 1   STRUCTURE AND FUNCTION OF THE BRAIN

To complete this activity, refer to **Figures 10-1 and 10-2.** You may also refer to human brain models, preserved brains, or other materials present in the classroom.

1. The brain is one of the body's most vital organs, and as such, needs protection. The first protective barrier consists of the bones of the skull. Although the skull does a good job of protecting the brain from external impact, internal cushioning is also required, since even a slight impact can cause the brain to hit the inside of the skull and result in a serious injury. To prevent this, the brain is cushioned and protected by three layers of connective tissue called the **meninges** (see **Figure 10-1).** Additional cushioning is provided by cerebrospinal **fluid between the layers.**

    Medical problems related to the brain and meninges are often mentioned in the news. For example, **meningitis** is an infection of the cerebrospinal fluid surrounding the brain or spinal cord. Meningitis is usually caused by a viral or bacterial infection that can damage or kill neurons and cause serious health problems.

Center for abstract thinking and problem solving, controls social behavior and expression of emotions. Controls voluntary motor activities and memories.

Frontal Lobe

Interprets sensations of touch, pressure, temperature, and taste. Coordinates information from a variety of senses.

Parietal Lobe

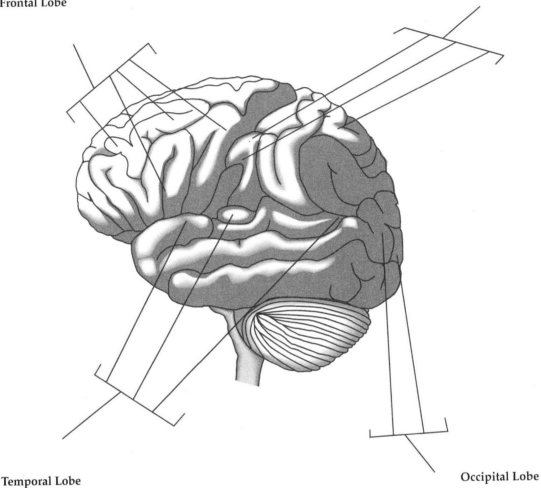

Temporal Lobe

Interprets and forms memories from auditory experiences. Functions in hearing, language skills, and understanding speech.

Occipital Lobe

Interprets visual images and coordinates vision with information from the other senses.

FIGURE 10-2.

2. The **cerebrum** (see **Figure 10-1**) is the largest part of the brain, covering almost the entire superior surface. It has two halves (called **hemispheres**), each controlling **the opposite side of the body.** The two halves are connected by a broad band of nerve tissues called the **corpus callosum.**

    What part of the body is affected if the **right side of the brain is injured** in an accident?

Observe the cerebrum and note its compact, folded structure. Within the body there are many different examples of ways to increase surface area and maximize functional capabilities. One of the most common ways is to **layer or fold** the tissue to fit into a smaller space. The cerebrum is an excellent example of this principle, since its highly folded surface is one of the most noticeable features of this part of the brain.

Each cerebral hemisphere is **divided into regions called lobes,** which control various aspects of conscious thought and problem solving abilities (see **Figure 10-2**).

3. Locate the **cerebellum**, a large, rounded structure positioned just below the occipital lobes of the cerebrum (see **Figures 10-1 and 10-2**).

The cerebellum processes information from the senses, joints, and muscles to coordinate movement of skeletal muscles, posture, balance, and equilibrium.

✓ Comprehension Check

Based on the diagram in **Figure 10-2**, fill in the blanks with the choice that is **most appropriate** to locate the part of the cerebrum that is most likely to be malfunctioning. Answers can be used **more than once.** Some questions may have **more than one correct answer.**

a. frontal lobe   c. parietal lobe
b. temporal lobe   d. occipital lobe

1. _____ Difficulty identifying colors.

2. _____ Difficulty understanding spoken words.

3. _____ Can't thread a needle.

4. _____ Doesn't feel pain when pricked with a pin.

5. _____ Noticeable changes in personality.

6. Since the cerebellum is related to coordinating movements, do you think the relative size of the cerebellum would be **larger or smaller in birds than in humans? Explain** your answer.

Check your answers with your instructor before you continue.

4. The **brain stem** (see **Figure 10-1**) connects the brain to the spinal cord. It is composed of three parts: **the midbrain, the pons, and the medulla oblongata.**

    All of the sensory and motor neurons carrying information to other areas of the brain pass through the brain stem. The brain stem sorts and directs information to the appropriate brain centers and acts as the main "highway" for nerve tracts traveling between the brain and spinal cord.

    Nuclei (localized control centers) within the **medulla oblongata** regulate several automatic body functions, such as heartbeat, breathing, and blood pressure.

5. There are four **ventricles** within the brain (see **Figure 10-1**). The ventricles form a series of **connected chambers** within the cerebrum and brain stem. The ventricles are filled with **cerebrospinal fluid** that circulates throughout the ventricles and the central canal of the spinal cord. The fluid circulates between the brain and spinal cord in the space between the meninges where it acts as a shock absorber, provides nutrients, and removes wastes.

6. Several additional internal structures can be identified on **Figure 10-1**. These include the thalamus, the hypothalamus, and the pituitary gland.

    The **thalamus** contains about a dozen specialized nuclei that sort and group nerve impulses by function and relay them to the appropriate regions of the cerebrum for action.

    The **hypothalamus** is one of the body's most important control centers for **homeostasis** (regulation of the internal environment). Among the important factors controlled by the hypothalamus are regulation of body temperature, hunger and thirst, blood sugar level, water balance, and emotional responses (such as sex drive).

    As part of its homeostatic function, the hypothalamus monitors body conditions, and responds by releasing **regulating factors**. Regulating factors are chemicals that stimulate or inhibit the release of hormones from the pituitary gland.

    The hypothalamus is linked to the **pituitary gland** (an endocrine gland located just beneath the hypothalamus). The pituitary gland produces a number of hormones that help to regulate growth, kidney function, reproductive cycles, milk production, and basal metabolic rate.

✓ Comprehension Check

1. Name **at least two** parts of the brain that would be activated when **riding a bicycle. Explain** your answer.

2. Name **five** parts of the brain that could be activated when you see an attractive women passing by in a string bikini.

Check your answers with your instructor before you continue.

ACTIVITY 2   SPINAL NERVES AND THE REFLEX ARC

The nervous system has several different types of nerve cells **(neurons). Sensory neurons** process and transmit incoming information to the central nervous system (the brain and spinal cord). **Motor neurons** carry information to muscles or glands, and the muscles or glands respond with the appropriate actions. **Interneurons** form **connections** between neurons.

Reflexes are unconscious, **involuntary motor responses.** A familiar spinal reflex occurs when you accidentally touch a hot stove. Before you even realize that your hand is painfully hot, a reflex has jerked your hand back from the stove.

During a **spinal reflex,** a sensory receptor is stimulated and the impulse is transmitted to the spinal cord. Many spinal reflexes are mediated only by the spinal cord and do not involve higher brain centers. This means that a motor neuron response can be stimulated by the spinal cord alone without input from the brain, although the brain can influence the strength of reflex responses.

Neurons vary in function and appearance, but all share a similar three-part structure: the cell body, the axon, and the dendrites. Every neuron has a **cell body,** the region of the cell that contains the nucleus and a variety of other organelles necessary for cell metabolism. The cell bodies of **most** neurons are located **within the central nervous system;** however, some are located **outside the CNS** in clusters called **ganglia.**

Outside stimuli are received by **dendrites** in the skin and other sense organs. Incoming signals from these dendrites are transmitted toward the cell body and continue into the **axon.** Axons relay outgoing messages from one neuron to other neurons or to various tissues and organs in the body.

1. Refer to the diagram in **Figure 10-3.**

   Label the following: **sensory neuron, interneuron, motor neuron, and the cell bodies of each of the three neurons.**

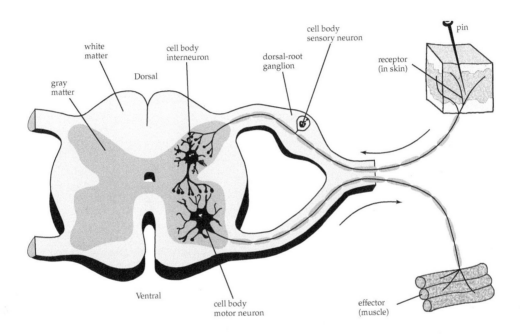

FIGURE 10-3.　Reflex Arc in the Spinal Cord

2. In which direction do impulses travel through sensory neurons?

   In which direction do impulses travel through motor neurons?

   (Circle one answer.) Sensory neurons / interneurons / motor neurons have the longest axons.

3. Work in **groups of two.** Get the following supplies: **two rubber reflex hammers.**

   Complete the following two activities to demonstrate the knee-jerk and plantar reflexes with your partner, then form a group of four students to complete the third reflex activity.

Knee-Jerk Reflex

1. Sit on a laboratory stool or table with your **legs relaxed and hanging freely. Make sure your feet do not touch the floor.**

2. With your finger, locate a soft area just below the patella (the knee cap).

    Gently tap this area with the reflex hammer and observe the response.

Note:

It may take you several tries to locate the exact spot to produce the patellar reflex response.

---

3. What were the results of your experiment?

4. **List, in order,** the events that occur in the nervous system during the knee-jerk reflex. Begin when the hammer hits the knee and conclude with the knee-jerk response.

5. **Exchange roles** and complete the plantar reflex activity.

Plantar Reflex (Negative Babinski Reflex)

1. Remove the shoe and sock from one foot.

    Sit on a laboratory stool and **support your foot** at the ankle on a second stool.

2. Gently draw the metal tip of the handle of the reflex hammer along the side of the sole **from the heel toward the big toe (across the ball of the foot).** Observe the response.

Blink Reflex

1. Work in **groups of four.** Get the following supplies: **clear plastic wrap and a piece of cotton.**

    Form the cotton into a large ball, about the size of a golf ball.

    Assign one member of your group to be the first test subject. A second group member will conduct the experiment, and the remaining members will observe and record the results of each trial.

    The test subject will hold the piece of plastic wrap in front of his/her face and eyes, far enough away from the face to serve as a barrier. The test subject should be **relaxed,** but attentive.

2. Throw the cotton ball at the plastic wrap. In **Table 10-1,** record if the test subject blinked or did not blink. **Repeat** the experiment **ten** times.

    Exchange roles and repeat the experiment until **each group member** has had an opportunity to be the test subject.

3. What percentage of trials produced blink responses for each test subject?

    Subject 1 _____ %

    Subject 2 _____ %

    Subject 3 _____ %

    Subject 4 _____ %

TABLE 10-1 BLINK REFLEX RESULTS				
TRIALS	BLINK (YES/NO) SUBJECT 1	BLINK (YES/NO) SUBJECT 2	BLINK (YES/NO) SUBJECT 3	BLINK (YES/NO) SUBJECT 4
1				
2				
3				
4				
5				
6				
7				
8				
9				
10				
TOTAL (Number of blinks)				

4. Since all the test subjects knew that the plastic wrap would prevent the cotton ball from hitting their faces, why did the blink response occur? **Explain** your answer.

✓ Comprehension Check

1. You're walking barefoot along the boardwalk at the beach, gawking at members of the opposite sex, when all of a sudden you step on a sharp object. The following activities will be occurring in your nervous system. Place the actions in the correct sequence.

_____ Muscles in your leg contract

_____ Sensory nerve endings are stimulated

_____ Brain senses pain

_____ Sensory neurons transmit impulses to the spinal cord

_____ Motor neurons transmit impulses to leg muscles

_____ Foot moves

_____ Members of opposite sex no longer seem so important

Check your answers with your instructor before you continue.

ACTIVITY 3   STRUCTURE AND FUNCTION OF THE SPINAL CORD

1. Get the following supplies: **a compound microscope and a prepared slide labeled "cross section of the spinal cord."**

   View the slide using the **scanning lens (4X)** of your microscope.

2. You'll note a "butterfly" shaped area with a darker stain towards the center of the spinal cord. This area of the spinal cord is composed **cell bodies** of interneurons and motor neurons. The tissue is known as **grey matter** because it appears grey to the naked eye.

   The white matter is composed of nerve fibers covered by a **myelin sheath,** which is wrapped around the axons of many neurons. When a signal travels along a myelinated axon, it can travel more than 300 miles per hour. Signals travel less than five miles per hour in axons without myelin sheaths.

3. **Spinal nerves,** which contain **nerve fibers** (bundled axons and dendrites), lie along the length of the spinal cord, but they can be difficult to see. Examples of spinal nerves can be seen in **Figure 10-3.**

    Refer to the figure. You can see that the spinal nerves divide into two pathways just prior to entering the spinal cord. These are called the **dorsal and ventral roots.** The **dorsal root** of a spinal nerve contains **sensory fibers** entering the gray matter of the spinal cord, and the **ventral root** of a spinal nerve contains **motor fibers** leaving the gray matter.

4. Focus your microscope on an area within the grey matter and switch up to **high power (40X).**

    Locate one or more neurons and observe the location of the following cell structures: **cell body, nucleus, cell membrane, axon, and dendrites.**

✔ Comprehension Check

Based on your microscopic observations and the diagram in **Figure 10-3,** answer the following questions.

1. **(Circle one answer.)**

    If an injury affected **only the dorsal root** of a spinal nerve, **sensory input/motor output** would be primarily affected.

    The cell body of the motor neuron is located in the **gray matter/white matter.**

    The dendrites of a motor neuron are located in the **grey matter/white matter/dorsal root/ventral root.**

    The cell body of a sensory neuron is located in the **grey matter/white matter/dorsal root/ventral root.**

2. **(Circle all that apply.)**

    The nerve fibers of **sensory neurons/motor neurons/interneurons** are located in spinal nerves.

    The myelinated axons of motor neurons are located in the **grey matter/white matter/dorsal root/ventral root.**

3. In anatomy, spinal nerves are sometimes classified as **"mixed nerves."** Why would this description be appropriate?

Check your answers with your instructor before you continue.

ACTIVITY 4   REACTION TIME

A reaction is a **conscious, voluntary response** to an outside event. As discussed in Activity 1, voluntary responses to stimuli are mediated by the cerebral cortex. You exert conscious control of your muscles when you chew food, manipulate your knife and fork, or take notes in class.

During a voluntary response, sensory receptors are stimulated and the impulse is transmitted through the spinal cord to higher brain centers. The incoming signal and response travel a greater distance than in a spinal reflex (which is mediated by the spinal cord. Therefore, a reaction response requires more time to complete than a reflex.

1. Work **with a partner.** Get the following supplies: **a reaction time ruler.**

2. Sit on a laboratory stool. You are the test subject. Your partner will stand next to you. He or she is the investigator conducting the reaction time experiment.

   The investigator will hold the meter stick vertically with the **release end** toward the **ceiling,** so that the **thumb line** of the ruler is approximately **chest height** on the seated **test subject.**

   The test subject should position the **thumb and fingers of one hand** around the meter stick at the location labeled **thumb line.** The thumb and fingers must be about **one inch away** from the sides of the ruler.

3. Once the investigator and test subject are correctly positioned, you can begin the experiment. **Without warning the test subject,** the investigator will release the meter stick and the test subject will catch the meter stick as quickly as possible. **Don't release your grip on the meter stick!**

    The investigator will observe the number on the stick closest to the **position of the test subject's thumb.** The ruler is calibrated such that the **position of the thumb** represents the test subject's **reaction time in milliseconds.** Record the results in **Table 10-2.**

4. Repeat the procedure nine more times, recording the results of each trial in **Table 10-2.** If time permits, switch places and repeat the experiment. **Calculate** the **mean** reaction time for each test subject.

TABLE 10-2 REACTION TIME RESULTS		
TRIALS	SUBJECT 1 RESPONSE TIME (msec)	SUBJECT 2 RESPONSE TIME (msec)
1		
2		
3		
4		
5		
6		
7		
8		
9		
10		
Mean reaction time		

5. **Calculate** the **mode** of your reaction times and determine the **range** of the data.

    mode = _____   range = _____

6. Suggest a method that could be used to determine whether your **average reaction time is faster or slower** than that of a typical college student.

Self Test

Choose the most appropriate answer for the questions below. Answers can be used **only once.**

a. ventral root   h. pituitary gland
b. cerebellum   i. white matter
c. corpus callosum   j. occipital lobe of cerebrum
d. grey matter   k. dorsal root ganglion
e. hypothalamus   l. meninges
f. central nervous system   m. thalamus
g. medulla oblongata   n. frontal lobe of cerebrum

1. _____ Integrates body position, motion, and balance.

2. _____ Interprets images seen on your computer screen.

3. _____ Includes the brain and the spinal cord.

4. _____ Control center for breathing, heart rate, and blood pressure.

5. _____ Area of the spinal cord that contains nerve fibers surrounded by a myelin sheath.

6. _____ Location of cell bodies of sensory neurons.

7. _____ Fibers of motor neurons pass through this part of the spinal cord.

8. _____ Connects the right and left cerebral hemispheres.

9. _____ Controls homeostatic functions, such as body temperature, thirst, and hunger.

10. _____ Center of consciousness and intelligence.

11. _____ Produces hormones that help to regulate growth and reproductive activities.

12. _____ Protective covering of the brain and spinal cord.

13.   Name a part of the brain that might be affected in an individual suffering from **Alzheimer's disease. Explain** your answer.

14.   Thousands of Americans are currently hospitalized in a persistent vegetative state (can only breathe and regulate other involuntary body functions on their own). What part of the brain is still functioning in these individuals? **Explain** your answer.

Introduction to Forensic Biology

1,2,4

Objectives

After completing this exercise, you should be able to:

- identify and interpret basic fingerprint patterns
- use your knowledge of fingerprint patterns and details to analyze forensic evidence
- demonstrate an understanding of the A–B–O and Rh blood typing systems
- use antigen–antibody interactions to identify and type unknown blood samples
- apply your knowledge of basic forensic techniques to real-life situations

CONTENT FOCUS

A scientist is similar to a detective. Both use scientific thinking to solve problems that have no clear solutions. Solving this type of open-ended problem requires you to develop your skills in forming hypotheses, gathering evidence by observation and experimentation, and drawing conclusions on the basis of your findings.

Forensic science is a good example of the practical application of the scientific method. In this case, science is used to analyze evidence and solve crimes. Forensic scientists use information from all branches of science, including chemistry, geology, physics, medicine, and, of course, **biology.**

This exercise provides an introduction to two of the basic techniques that are employed in forensic biology: **fingerprint analysis and blood typing.**

ACTIVITY 1   GETTING STARTED

Fingerprinting is the most commonly used and reliable system of identification. Each person has a unique pattern of ridges and valleys on his/her fingertips. Although other physical characteristics (such as weight, height, and hair color) may change as a person ages, fingerprints remain the same for life. Fingerprints penetrate through five skin layers and so are almost impossible to erase. Criminals have tried filing, acid burning, and even surgical removal, but with limited success.

Fingerprints have been used for identification purposes for several thousand years. The Babylonians recorded their fingerprints in the soft clay of their writing tablets (paper wasn't yet in use). The fingerprints served as a "signature" to prevent document forgeries. A similar system was also used in ancient China and Japan. In the last 100 years, an internationally recognized system of identifying fingerprints has been developed. Most law enforcement agencies (even in local communities) maintain their own fingerprint files. They can also access fingerprint information from state agencies and the FBI.

The skin is covered with **sweat glands,** including some in the ridges on the fingers and palms. During normal activities, perspiration from the sweat glands accumulates in these ridges. These ridges also accumulate body oils (from touching oil-producing areas such as the face, scalp, or neck). When you touch an object, perspiration and skin oils are transferred to that surface, leaving an invisible impression (the **latent fingerprint**). Latent prints found on nonabsorbent surfaces (such as metal or glass) can be dusted with colored powder and removed with transparent tape. Latent prints on absorbent surfaces (such as cloth or paper) must be recovered chemically.

The **Fingerprint Classification System** has three basic patterns: **arches, loops, and whorls.**

1. Begin by learning to recognize the basic identifying patterns used in fingerprint analysis.

Pattern area	Part of the fingerprint that is used for identification **(Figure 11-1a)**
Ridgelines	Individual lines that make up a fingerprint
Delta	Area where **two ridgelines** come to a **point (Figure 11-1b)**
Core	Approximate center of the finger impression **(Figure 11-1b)**
Whorl	**Complete circle** formed by ridgelines between two deltas; whorls **don't** continue off either side of the print **(Figure 11-2)**

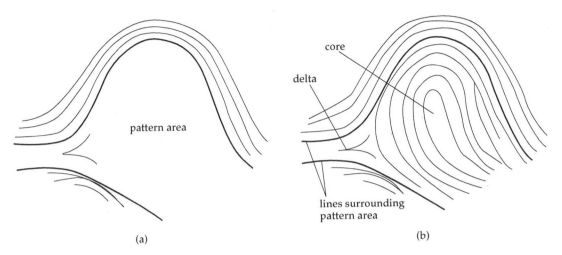

FIGURE 11-1. (a) Fingerprint Pattern Area and (b) Features Used in Identifying Fingerprints

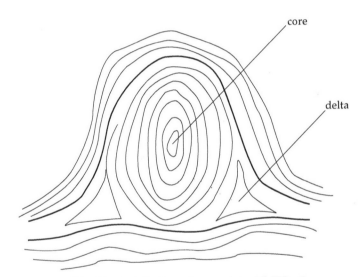

FIGURE 11-2. Fingerprint with Whorl

✓ Comprehension Check

1. Refer to **Figure 11-1b.** How many **ridgelines** are there between the delta and the core? __5__

2. **(Circle one answer.)** The fingerprint in **Figure 11-1b** is / is not a whorl.

3. Refer to **Figure 11-2.** How many **ridgelines** are there between the delta and the core? __8__

Check your answers with your instructor before you continue.

2.   In addition to whorls, there are several other common fingerprint patterns.

Simple arch	Ridgelines **enter** on the **left,** rise, and **exit** on the **right** of the print (**Figure 11-3a**)
Tented arch	Ridges meet at the center and form a **peak** (**Figure 11-3b**)

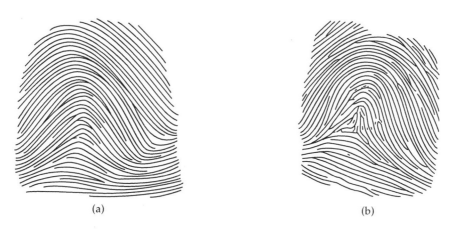

(a)   (b)

FIGURE 11-3.   (a) Simple Arch and (b) Tented Arch

3.   There are various types of loops.

Loops	Curved ridgelines that **enter** and **exit** on the **same side** of the print (**Figure 11-4a**); loops can be either **left or right**-facing
Double loop	Two separate loop formations on the same finger (**Figure 11-4b**)

(a)   (b)

FIGURE 11-4.   (a) Loop and (b) Double Loop

4. The fingerprint classification system doesn't permit designating a fingerprint as more than one type. **For example, you can't designate the same fingerprint as both a loop and an arch.** Look at the fingerprint and identify the **largest, most dominant pattern.** This is the pattern you'll use to designate the print.

    Since you can't apply more than one fingerprint pattern to a single print, fingerprints that don't fit easily into any of the three categories are sometimes **grouped into a fourth category,** called **mixed.** Mixed prints usually show a combination of two or more different patterns.

5. Since there are only a few possible patterns, additional identifying characteristics are required to make a fingerprint match. A few examples are listed below:

Bifurcation	Forking or dividing of one ridgeline into two or more branches **(Figure 11-5)**
Divergence	Spreading apart of two ridgelines that have been running parallel or nearly parallel **(Figure 11-5)**

    Since scars and other marks that accumulate on your fingers as a result of daily wear and tear are unique, they are convenient features to use for identification. **Scars appear as white lines** on the fingerprint.

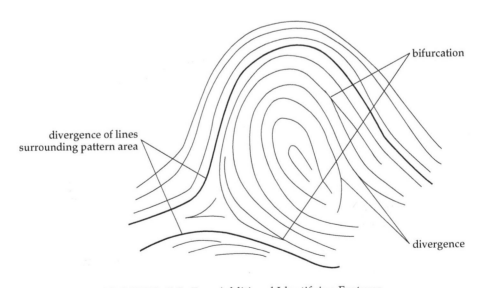

FIGURE 11-5. Additional Identifying Features

✔ Comprehension Check

1. Place an "**X**" in front of each feature that is found in the fingerprints in **Figure 11-6.**

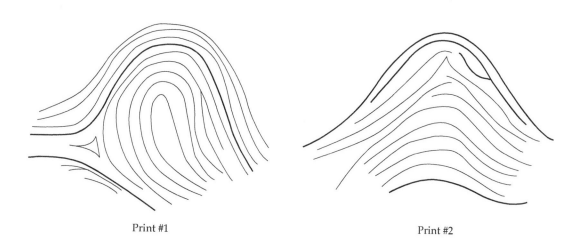

Print #1   Print #2

X	Characteristic
✗	Delta
✗	Whorl
	Simple arch
	Tented arch
✗	Loop
	Double loop
	Mixed
	Divergence of ridgelines
✗	Bifurcated ridgelines

X	Characteristic
✗	Delta
	Whorl
✗	Simple arch
	Tented arch
✗	Loop
	Double loop
	Mixed
✗	Divergence of ridgelines
	Bifurcated ridgelines

FIGURE 11-6. Analysis of Fingerprints

Check your answers with your instructor before you continue.

ACTIVITY 2

THE FINGERPRINT "FORMULA"

A fingerprint formula is a list of **print classifications for each hand.** Each hand has a distinctive fingerprint formula. Before you can analyze crime-scene prints, you must develop some expertise in recognizing the different fingerprint patterns.

1. Work in groups. Get the following supplies: **one fingerprint inkpad per group, magnifying glasses, and some paper towels.**

2. Remove **Figure 11-7** from your lab book and place it on the table in front of you. Using the fingerprint inkpad, make a clean set of fingerprints for each member of your team.

    Place the prints on the **Fingerprint File Card** in **Figure 11-7.** To make the prints, place **one side** of each finger on the inkpad. **ROLL** the finger from one side to the other, covering the **FRONT** (**NOT** the tip) with ink.

Note:

Be careful with the fingerprint ink. It can stain your clothes.

---

3. Repeat the rolling motion to transfer the inked print onto **Figure 11-7** (or a clean sheet of white paper).

    If you press too hard, you'll get a black ink blob with no discernable ridges. Use a light touch to obtain a readable print.

Left Hand   Thumb

Right Hand

Thumb

FIGURE 11-7. Fingerprint File Card

4. To prepare an individual's fingerprint formula, **begin at the thumb and proceed to the little finger.** For example, a hand that has fingers of **arch, arch, whorl, loop, whorl** has a print formula of **a–a–w–l–w.**

   Using your own fingerprints, **determine your fingerprint formula.** Write the fingerprint formula **directly underneath each fingerprint in Figure 11-7.**

5. **By group discussion, double-check** each person's assessment of their own fingerprint formula.

   Hard-to-classify prints may require your group to clarify and add information to the definitions of each category. As the distinctions between print patterns become more clearly defined, it'll be easier for you to assign individual prints to a print type.

Check your answers with your instructor before you continue.

6. Using your revised print type descriptions, **classify the twelve fingerprints in Figure 11-8.**

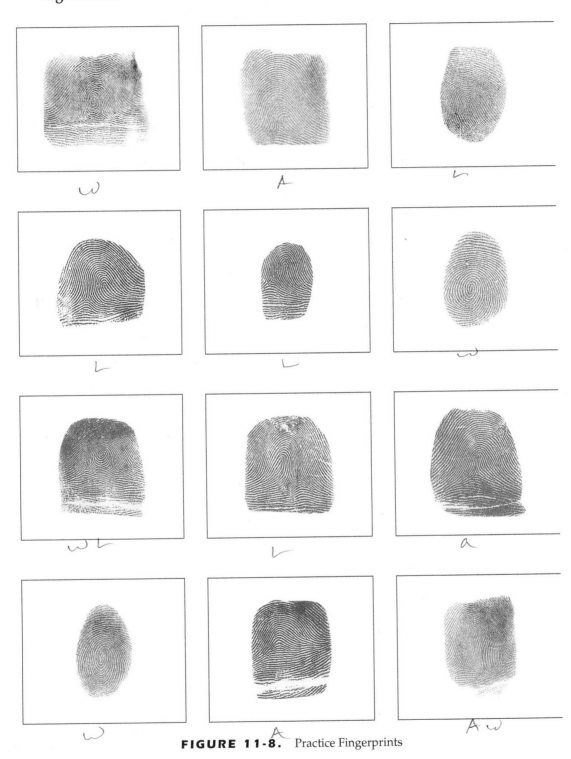

FIGURE 11-8. Practice Fingerprints

Check your answers with your instructor before you continue.

ACTIVITY 3 DETERMINING POINTS OF SIMILARITY

Since there are only a few possible fingerprints, it's necessary to look at the **fine details of the ridgelines** within the prints (such as bifurcations, divergences, etc.) to match sample fingerprints to a specific person.

Exact matches in ridgeline patterns between two prints are called **points of similarity.** For a conclusive match between two prints, a minimum of six points of similarity is usually required. Since you're just learning how to do this, **one or two points of similarity per fingerprint** will be sufficient.

Circle each point of similarity you find, mark it with a letter designation, and provide an explanation for each point, as shown in **Figure 11-9.**

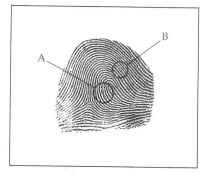 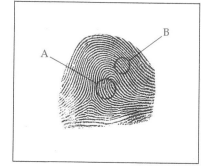

Sample print 1   Sample print 2

Explanation

Both prints are right-facing loops

A   Bifurcation forming a broken oval pattern with a complete oval pattern below

B   Oval pattern in ridgeline

FIGURE 11-9.   How to Label Points of Similarity

✓ Comprehension Check

Mr. Akumba arrives home and discovers that his back porch window has been forced open. He looks around the house and discovers that his new TV is missing. When the police arrive, they dust the scene for fingerprints. A few days later, Ms. Wright is arrested while breaking and entering another house in the neighborhood. You're the fingerprint examiner for the police department. The chief detective has asked you to compare Ms. Wright's fingerprints with those found at Mr. Akumba's house (**Figures 11-10** and **11-11**).

Left Hand   Thumb

Right Hand

Thumb

FIGURE 11-10. Fingerprint File Card for Ms. Wright

FIGURE 11-11. Prints Lifted from Mr. Akumba's Window

1. Do any of the prints from the crime scene match those of the suspect? If so, which prints match which fingers? List the matches below.

2. List the **points of similarity** you used to make your match. Mark each point with a letter on the fingerprint and provide an explanation for each letter (as shown in **Figure 11-9**).

3. **Did Ms. Wright do wrong?** Were those prints at Mr. Akumba's house hers? In a few sentences, **summarize the conclusion** you reached on the basis of your fingerprint examination.

Check your answers with your instructor before you continue.

ACTIVITY 4   BLOOD TYPING

Scientists have identified and studied many different human blood groups. Blood groupings are based on the presence of **identifying proteins (also called antigens)** that are incorporated into the cell membranes of all body cells. Although red blood cells have **no personal identifying proteins,** they do have antigens that give us our different blood types. These molecules are referred to as antigens because they are identified as foreign if placed in the body of a person who doesn't have these molecules.

The most commonly known antigens are those that determine the **A–B–O and Rh** blood groups. These blood groups are well known because they are used to type blood for transfusions and organ transplants. The groups include the following:

- **Type A** blood contains antigen **A** and forms antibodies against **B.**

- **Type B** blood contains antigen **B** and forms antibodies against **A.**

- **Type AB** blood contains both antigens **A and B.** Type AB **doesn't form antibodies against either A or B.**

- **Type O** doesn't contain **A or B antigens,** but forms antibodies against **both A and B.**

- The **Rh factor** is the name given to another antigen present in red blood cells. People with **Rh+ blood have the antigen,** while those with **Rh− blood don't.**

Naturally occurring **antibodies (defensive proteins)** in the blood cause a serious transfusion reaction called **agglutination** if a person is transfused with blood containing **foreign** antigens of the A–B–O or Rh blood groups. Babies aren't born with these specialized antibodies, but they begin to develop them a few months after birth.

The combination of an antigen and its corresponding antibody produces **agglutination (clumping of blood cells).** The agglutination reaction makes it possible to determine the blood type.

To determine a person's blood type, we'll cause agglutination to happen in the laboratory. Of course, we can't do this test in a person's bloodstream, because that could cause harmful agglutination within the blood vessels. However, we can place the blood samples in the wells of a depression plate and simulate the reaction, so that you can observe how the clumping takes place.

The antibodies that bind to the blood antigens are labeled according to the antigen they bind to. For example, **Antibody-A (which we'll refer to as "anti-A" binds to and causes clumping in type A blood (because type A blood contains the matching type A antigen).**

> **Note:**
>
> If we add anti-A to type A blood, clumping will occur. The addition of antibody-B, however, wouldn't cause these cells to clump.

By analyzing the reactions of an unknown blood sample with specific antibodies, and comparing your results to those in **Table 11-1,** the blood type can be determined.

1. Work in groups. Get the following supplies: **a spot plate, a wax pencil, some toothpicks, a magnifying glass, and dropper bottles of the following: known samples 1 and 2, unknown samples 1 and 2, anti-A serum, anti-B serum, and anti-Rh serum.**

> **Note:**
>
> These tests are being carried out with simulated blood, which is nonbiological and nontoxic.

TABLE 11-1
CLUMPING REACTIONS THAT OCCUR IN BLOOD TYPING

BLOOD TYPE	EFFECT OF ADDING ANTIBODY-A	EFFECT OF ADDING ANTIBODY-B
A	Clumps form	No clumps
B	No clumps	Clumps form
AB	Clumps form	Clumps form
O	No clumps	No clumps

2. You'll be testing each blood sample for the presence of the antigens A, B, and Rh.

    With the wax pencil, label **three horizontal rows** of your spot plate **A, B, and Rh** as shown in the sample in **Figure 11-12.**

3. You'll be practicing on two blood samples of known type **(K1 and K2).** With these two samples as **controls,** you'll be able to see exactly what the clumping patterns look like for each possibility of A, B, and Rh antigens. Your instructor will let you know the blood types of the known samples.

    Fill in the blanks next to K1 and K2 in the first two vertical columns of your spot plate with the **names of the blood types provided by your instructor.**

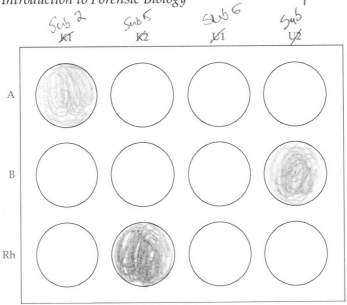

FIGURE 11-12. Sample Spot Plate

4. Place one large drop of **anti-A** solution into **depressions K1 and K2** in the row labeled **A.** Place one large drop of **anti-B** solution into **depressions K1 and K2** in the row labeled **B.**

   Place one large **drop of anti-Rh solution** into **depressions K1 and K2** in the row labeled **Rh.**

5. For each of the three depressions for the known **sample K1,** put a drop of blood from the appropriate dropper bottle.

   Stir each sample with a toothpick. **Use a CLEAN TOOTHPICK for each sample.**

6. For each of the three depressions for known **sample K2,** put a drop of blood from the appropriate dropper bottle.

   Stir each sample with a toothpick. **Use a CLEAN TOOTHPICK for each sample.**

7. Observe the mixtures in the six depressions.

   If the agglutination test is **positive,** you'll see a **collection of small clumps,** or particles, in the depression.

   You may find this easier to see by using a magnifying glass or dissecting microscope.

Note:

Don't empty the liquid from the depressions containing the controls.

8. Return to the diagram in **Figure 11-12.** With a colored pencil or a highlighter, **color** each depression of the **known samples** where you observed a **clumping** reaction.

9. Repeat the same test procedures using the two unknown blood samples (U1 and U2) supplied by your instructor.

10. Return to the diagram in **Figure 11-12.** With a colored pencil or a highlighter, **color** each depression of the **unknown samples** where you observed a **clumping** reaction.

11. The blood type of unknown **sample U1** is ___A+___.

    The blood type of unknown **sample U2** is ___B+___.

    Explain your answers.

✓ Comprehension Check

Owing to a clerical error, several samples of blood stored at the local blood bank may be incorrectly labeled as to the blood type. The three tests listed below were conducted on each sample. Use your knowledge of antigen–antibody reactions to sort the samples into their correct blood types.

Test 1   Unknown sample mixed with "anti-A" serum (contains "type A" antibodies)

Test 2   Unknown sample mixed with "anti-B" serum (contains "type B" antibodies)

Test 3   Unknown sample mixed with "anti-Rh" serum (contains antibodies against the Rh protein)

1. **Unknown Sample A:** When tested, agglutination (clumping) of red blood cells occurred in Tests 1, 2, and 3.

   Blood Type: _____

2. **Unknown Sample B:** When tested, no agglutination occurred in any of the three tests.

   Blood Type: _____

3. **Unknown Sample C:** When tested, agglutination occurred in Tests 2 and 3, but not with Test 1.

   Blood Type: _____

4. **In your OWN words,** explain the chemical reaction that occurs when anti-B serum is mixed with

   a. a type A blood sample:

   b. a type B blood sample:

c. an Rh-positive blood sample:

Check your answers with your instructor before you continue.

Note:

Blood typing can be useful; however, it isn't highly informative. Since there are only four blood groups in the A–B–O system, many people share the same type. Even if the blood type of the evidence matches the suspect, it doesn't prove that this blood came from the suspect. It could be from anyone that has the same blood type. For this reason, forensic scientists have moved away from conventional blood typing toward the more specific DNA typing technology.

Self Test

Fill in the blanks with the choice that is **most appropriate. Answers can be used only once.**

a. ridgeline
b. delta
c. whorl
d. tented arch
e. loop

f. bifurcation
g. divergence
h. fingerprint formula
i. points of similarity
j. sweat glands

1. _____ Complete circle formed by ridgelines between two deltas.

2. _____ Two parallel ridgelines spread apart.

3. _____ Ridgeline divides into two or more branches.

4. _____ Curved ridgelines that enter and exit on the same side of the print.

5. _____ Produce fluids that form latent fingerprints.

6. _____ Set of print classifications for each hand.

7. _____ Exact matches of ridgeline patterns between two fingerprints.

8. _____ Fingerprint pattern in which ridges meet at the center to form a peak.

9. Why do your fingers leave prints when you touch something?

10. Why is it unlikely that identical twins would have identical fingerprints?

For questions 11–13, identify the most likely blood type on the basis of the results of the following antigen–antibody reactions. Explain each answer.

11.   When tested, agglutination occurred when exposed to anti-A serum, but not with anti-B or anti-Rh serums.

12.   When tested, agglutination occurred when exposed to anti-A serum, and also with anti-B serum. No agglutination occurred with anti-Rh serum.

13.   When tested, agglutination didn't occur when exposed to both anti-A and anti-B serums. Agglutination did occur with anti-Rh serum.

The Circulatory System

Objectives

After completing this exercise, you should be able to:

- diagram the pattern of blood circulation around the body and label the blood in each location as oxygenated or deoxygenated
- differentiate between the pulmonary and systemic circuits
- identify and explain the function of each of the major structures of the heart
- discuss the connection between blood vessel diameter and the rate of blood flow
- discuss the effect of exercise on pulse rate, blood pressure, and the efficiency of oxygen transport
- discuss the ways in which exercise changes the distribution of blood in the body
- apply your knowledge of vessel and heart structure to cardiovascular health issues

CONTENT FOCUS

We hear a lot about exercising our muscles and keeping fit. Activities that increase cardiovascular fitness are an important part of any exercise program. Many people don't realize that the heart is a muscular pump, and so, as with all our body muscles, heart function can be improved by exercise and activity.

The average heart is only about the size of your fist. It normally beats about **72 times per minute.** This may not sound too impressive, but each day, the heart pumps **1500 gallons** of blood. Over your lifetime, this adds up to enough blood to fill **13 supertankers!**

Each beat of the heart **transports blood** that contains **food, oxygen,** and **hormones** to all the cells of the body and **removes wastes** such as **carbon dioxide.**

The circulatory system of humans and most other animals consists of **three basic elements: a pump, blood vessels, and blood (the circulatory fluid).** In this exercise, you'll take a closer look at the heart and do some experiments to demonstrate how the circulatory system adjusts to the needs of the body.

ACTIVITY 1

CIRCULATION OF BLOOD AROUND THE BODY

The heart is divided into **four** chambers. The two **upper** chambers, called **atria,** receive blood returning to the heart. The two **lower** chambers, called **ventricles,** pump blood out.

1. The circulatory pathway that carries blood through the lungs and back to the heart is called the **pulmonary circulation.** The circulatory pathway that carries blood through the upper and lower body and back to the heart is called the **systemic circulation.** Following these pathways, you can see that **blood returns to the heart TWICE as it circulates around the body.**

    As shown in **Figure 12-1, blood returns from the body** into the **right atrium.** The blood then travels into the **right ventricle,** which pumps it out to the **lungs.** Blood **returns from the lungs** into the **left atrium,** and then goes into the **left ventricle,** which pumps it to the **rest of the body.**

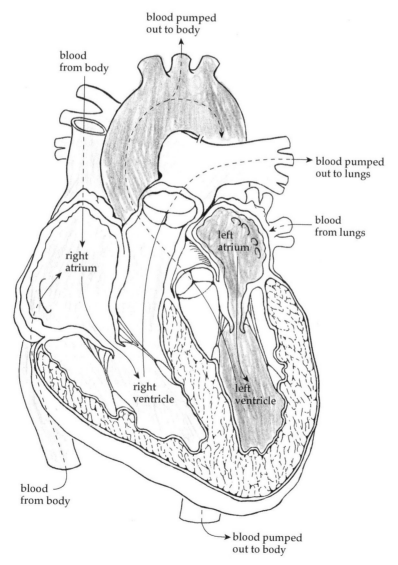

FIGURE 12-1. Circulation of Blood Around the Body

2. Using a **colored pencil or highlighter** (and the information you've studied about lung function), color each **heart chamber in Figure 12-1** that contains **deoxygenated** blood (blood **low** in oxygen) in **blue.** Color each **heart chamber** that contains **oxygenated** blood (blood **high** in oxygen) in **red.**

3. Within the pulmonary and systemic circulatory systems, the blood travels through **arteries, veins,** and **capillaries.**

   Arteries always transport blood **away from the heart. Veins** carry blood **toward the heart. Between the arteries and veins** lie **beds of capillaries** where cells pick up oxygen and release carbon dioxide.

   On **Figure 12-2, place a letter "A" next to** each blood vessel that is an **artery.** Place a **letter "V"** next to each **vein.** Place a **letter "C"** next to each **capillary bed.**

4. On **Figure 12-2,** color each **heart chamber and blood vessel** that contains **deoxygenated** blood in **blue.**

   Color each **heart chamber and blood vessel** that contains **oxygenated** blood in **red.**

5. Using the **clues** below, correctly **label** the listed blood vessels on **Figure 12-2.**

BLOOD VESSEL	CLUE FOR IDENTIFICATION
Superior vena cava	Vessel that returns blood from the head to the heart
Inferior vena cava	Vessel that returns blood from the lower body to the heart
Pulmonary artery	Vessel that carries blood to the lungs
Pulmonary vein	Vessel that returns blood from the lungs to the heart
Aorta	Vessel that distributes oxygenated blood around the body
Carotid artery	Branch off the aorta that delivers blood to the head

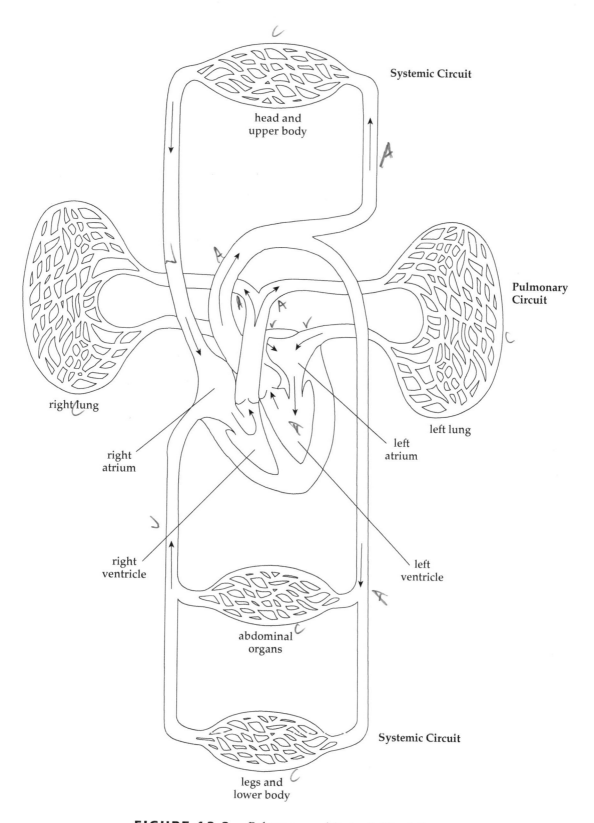

FIGURE 12-2. Pulmonary and Systemic Circulation

✓ Comprehension Check

1. **(Circle one answer.)** A drop of blood in the **pulmonary artery** is **oxygenated /
   deoxygenated.**

   Which heart chamber did it come from? _____

   Where will this drop of blood be found next? _____

2. **(Circle one answer.)** The blood flowing through the **left side** of the heart is
   oxygenated / deoxygenated. Explain your answer.

3. **(Circle one answer.)** A drop of blood in the **inferior vena cava is oxygenated /
   deoxygenated.**

   Which heart chamber will it enter next? _____

   Where will this drop of blood travel next? _____

4. Do all arteries carry **oxygenated blood? Explain** your answer.

**Check your answers and your Figure 12-2 labels with your instructor
before you continue.**

ACTIVITY 2   A CLOSER LOOK AT THE HEART

1. Work in groups. Get a **model of the human heart.** Using **Figure 12-3** as a guide, locate the **four chambers** on the heart model. **Observe** the muscular walls of the left and right ventricles on the heart model.

    On which side of the heart is the muscular wall the **thickest?** *left*
    Why? *Where it pumps the hardest*

Hint:
Review the information in Figures 12-2 and 12-3.

2. **Heart valves** prevent blood from flowing **backward** through the heart. Each valve section is anchored to the heart wall by **tendinous cords.** These strong cords **tighten** to hold the **valve sections closed** when the ventricles contract. When the cords are **relaxed,** the valve **opens.**

    On your model of the heart, locate **one valve between the right atrium and the right ventricle** (the **tricuspid valve**) and a **second valve** between the **left atrium and the left ventricle** (the **bicuspid** or **mitral valve**).

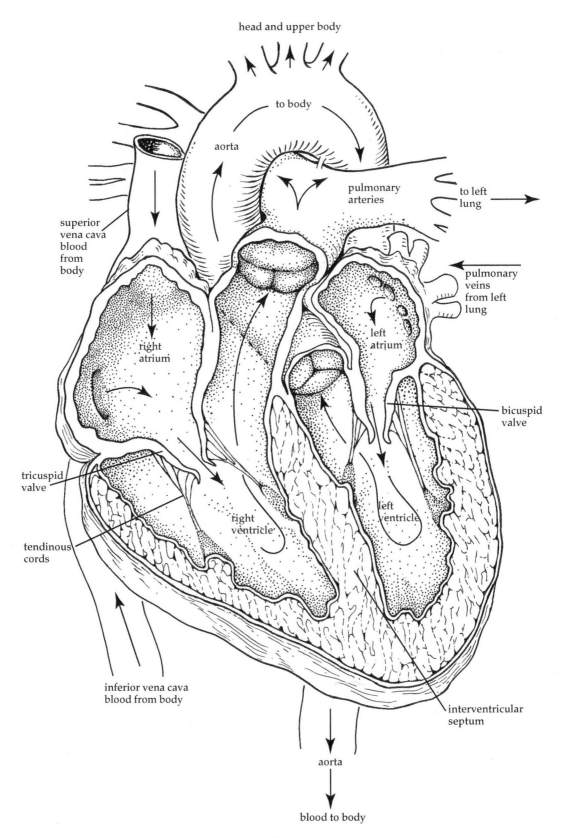

FIGURE 12-3. Interior View of the Heart

3.   To beat properly, the muscles of the heart must get an adequate blood flow.

Look at the **outer surface** of the heart model. You'll see many **small arteries and veins** branching across the surface of the heart.

The blood that flows through the heart is moving too fast to be used as a source of oxygen and nutrients by cardiac muscle cells.

A branch off the **aorta** connects to the **coronary arteries,** which lead to capillary beds in the heart muscle. Blood is returned through the **coronary veins** to the inferior vena cava.

✔ Comprehension Check

1.   **Circle the correct answer** in each of the following statements.

a. The **bicuspid** valve is **opened / closed** when the **left atrium** contracts.

b. The **tricuspid** valve is **opened / closed** when the **right ventricle** contracts.

c. The **tendinous cords** of the **tricuspid** valve are **tightened / relaxed** when the right ventricle contracts.

2.   Blood that leaves each ventricle is pumped into an artery. Another valve prevents blood from flowing backward into the ventricle. These are called the **pulmonary and the aortic semilunar valves. Locate** the **semilunar valves** on the heart model (see **Figure 12-4**).

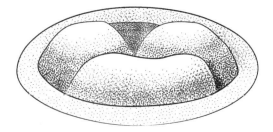

FIGURE 12-4.   Semilunar Valve

3. The semilunar valve on the **right** side of the heart is between **which heart chamber and which blood vessel?**

    _____   _____

4. The semilunar valve on the **left** side of the heart is between **which heart chamber and which blood vessel?**

    _____   _____

5. **(Circle one answer.)**

    a. When the right **ventricle** is **contracting,** the **pulmonary** semilunar valve is **opened / closed.**

    b. If a drop of blood is moving **toward the lungs,** it has passed through the **pulmonary / aortic** semilunar valve.

6. **Challenge Question!** When the right **atrium** is **contracting,** the **pulmonary** semilunar valve is **opened / closed.**

Check your answers with your instructor before you continue.

ACTIVITY 3 — BLOOD VESSEL DIAMETER AND BLOOD FLOW

The ventricles contract to pump blood through all the arteries of the body. As blood travels away from the heart, it passes through smaller and smaller blood vessels.

How hard does the heart have to work to pump blood through blood vessels of different sizes?

How much force the heart exerts to pump blood **depends on how easily fluid passes through the blood vessel.** Using **rubber tubing to simulate blood vessels,** you'll design an experiment to determine whether the **diameter of blood vessels** affects the **rate of blood flow.**

During your experiment, you want to consider **only** the effect of blood vessel diameter. For this reason, you **won't** use a pump to simulate heart action. Blood will flow through your simulated blood vessels using only the force of **gravity.**

1. Work in groups.

   Get the following supplies: **a large container with spigot, a stopwatch, a plastic bucket, a large graduated cylinder, and three pieces of plastic tubing** (see sizes below):

   Large   6/16″ inside diameter

   Medium   5/16″ inside diameter

   Small   4/16″ inside diameter

2. **Develop a hypothesis** about the relationship between the **blood vessel diameter** and the **rate of blood flow.** Write it in the space below.

BLOOD FLOW HYPOTHESIS

Check your hypothesis with your instructor before you continue.

3. Carefully **develop a plan** about how you'll get the information you need to test your hypothesis. **List the steps** of your plan, **including** the **equipment** you intend to use.

STEPS OF YOUR EXPERIMENTAL PLAN

4. In the space below, **make a table or chart** that shows your results clearly and neatly. Collect your experimental data and **record the results.**

5. **Plot a graph of your results in Figure 12-5.**

FIGURE 12-5. _____

6. Discuss the results with your group members. Write a couple of sentences that **summarize** your **results.**

Hint:
Remember — results include only facts, never opinions.

---

7. **Write a conclusion based on your hypothesis and collected data. Support your conclusion** by mentioning facts collected during your experiment.

8. **(Circle one answer.)** As blood travels from arteries into arterioles, the rate of blood flow will **increase / decrease.**

9. Capillaries are the **smallest** blood vessels. How does this fact affect the **rate** of blood flow?

10. Why is **flow rate** an important factor for determining how **efficiently** cells can pick up oxygen and remove carbon dioxide?

11. In the disease **atherosclerosis,** fatty plaques accumulate on the inside of artery walls. What effect would this have on the **rate** of blood flow? **Explain** your answer.

 Check your answers with your instructor before you continue.

ACTIVITY 4   THE EFFECT OF EXERCISE

Every time the ventricles contract **(systole),** forcing blood out of the heart, pressure in the blood vessels **increases.** When the ventricles **relax (diastole),** pressure in the blood vessels **drops** again. The pressure exerted by the blood on the walls of the blood vessels is called **blood pressure.** In the original instrument used to measure blood pressure, a **column of mercury** was forced upward in a glass tube. Newer instruments that measure blood pressure still use the same measurements (**millimeters of mercury** or **mm Hg**). The average blood pressure in a young adult male is approximately **120 mm Hg during systole** and **80 mm Hg during diastole.** This is expressed as **120/80.** In young adult females, the average is about 8–10 mm Hg less than in males. Blood pressure can be lower in adults who exercise regularly.

Blood is forced into your arteries during each contraction of the heart. You can feel this wave of blood moving through the arteries as a **pulse** in the carotid artery of your neck. The pulse can also be felt in the radial artery of your wrist and other locations in the body. The **number of pulses** tells you **how fast the heart is beating.** The average pulse rate in a young adult is **72 beats per minute.**

1. Work in **groups.** Do the following experiments.

   One student will be the test subject. Others will monitor blood pressure and pulse rate, serve as timekeepers, and record the experimental data.

Caution!
DON'T be the test subject for this activity if you have heart or blood pressure problems!

2. Get the following supplies: **a blood pressure monitor, a stopwatch, and an aerobic step.**

3. Sit quietly for **one minute. While sitting,** take your **resting pulse rate and blood pressure.**

    Resting pulse rate: _____79_____

    Resting blood pressure: __140/60_____

4. **Rapidly step up and down** on the aerobic step for **five minutes.** You should really be working out!

    Immediately sit down and record the **pulse rate and blood pressure.**

    Exercise pulse rate: _____110_____

    Exercise blood pressure: __175/85_____

5. Is there a difference in pulse rate before and after exercise? __Yes_____

    If so, **describe the difference.**

    higher because the heart is pumping more

6. Is there a difference in blood pressure before and after exercise? __Yes____

    If so, **describe the difference.**

    The hart is moving more blood.

7. What is the heart doing when **blood pressure** increases?

    Puumping more.

    What is the heart doing when the **pulse rate** increases?

    Pumping faster.

8. How do these **two different changes in heart action** help the body during exercise? In your answer, use the following terms: **blood flow, oxygen, carbon dioxide, cell respiration, and ATP.**

Increasing the blood flow, and oxygen is good for your heart. and needs ATP for cell regeneration. also when your heart is pumping fast it pushes out carbon dioxde

Self Test

1. Follow a drop of blood around the body. Begin with the tissue capillaries supplying your left big toe. **Place these locations in the correct sequence.**

    a. __1__ Tissue capillaries in big toe

    b. _____ Pulmonary artery

    c. _____ Left atrium

    d. _____ Pulmonary vein

    e. _____ Right atrium

    f. _____ Left ventricle

    g. _____ Inferior vena cava

    h. _____ Aorta

    i. _____ Right ventricle

    j. _____ Arterioles (small arteries)

    k. _____ Venules (small veins)

    l. _____ Lungs

2. If there were **no tendinous cords** attached to the tricuspid valve, and the right ventricle is contracted, where would the blood go?

3. If a person's pulse rate is 96 pulses per minute, what is this person's **heart rate?**

4. For the following body locations, enter **"D"** if the blood is **deoxygenated** and **"O"** if the blood is **oxygenated.**

    _____ Tissue capillaries entering big toe

    _____ Pulmonary artery

    _____ Left atrium

    _____ Pulmonary vein

    _____ Right atrium

    _____ Left ventricle

    _____ Inferior vena cava

    _____ Aorta

    _____ Right ventricle

    _____ Tissue capillaries leaving big toe

    _____ Arterioles

    _____ Venules

5. You're a doctor listening to heart sounds through a stethoscope. While listening, you notice an unusual "hissing" sound in a patient's heart. On consulting the patient's medical records, you find that as a child he suffered from rheumatic fever, which causes scar tissue to form around the heart valves. What might be causing the hissing sound?

6. When a person **blushes,** blood flow to the skin increases. In order for this change in blood flow pattern to occur, how must the **diameter** of skin arteries change?

7. Place an "**X**" in front of the **one** measurement of oxygen concentration **that's most likely to be correct for all four locations.**

X	Right Ventricle	Pulmonary Artery	Pulmonary Vein	Left Atrium
	40	40	100	100
	100	40	40	100
	100	100	100	100
	40	40	40	40
	40	100	40	100

The graph in **Figure 12-6** shows the distribution of blood (by percentage) in the circulatory system when the body is at rest and during heavy exercise.

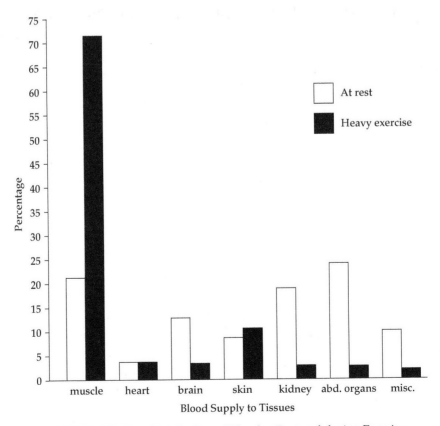

FIGURE 12-6. Distribution of Blood at Rest and during Exercise

8. In reference to the graph in **Figure 12-6,** what **change** occurs to the **percentage of blood supply to the muscles** when you're exercising?

   Considering the **change in the percentage of blood flow** to the muscles shown in the graph, what probably happens to the **diameter of the blood vessels** that supply blood to the muscle tissue during exercise?

9. What **change** occurs in the percentage of blood supply to the **abdominal organs** during exercise?

   Considering the **change in the percentage of blood flow** to the abdominal organs shown in the graph, what probably happens to the **diameter of the blood vessels** that supply blood to the abdominal organs during exercise?

10. What **change** occurs in the **percentage of blood supply to the skin?**

    How is this **change** in blood flow **helpful to the body during exercise?**

Introduction to Anatomy: Dissecting the Fetal Pig

1-5

Objectives

After completing this exercise, you should be able to:

- identify and compare the external anatomical features of the male and female fetal pig
- identify and explain the function of each of the major structures of the mouth, the neck, and the thoracic cavity
- explain how organ structures are specialized to perform specific functions and give examples
- compare and contrast anatomical features of the pig with those of humans
- discuss the difference between lung capacity during normal breathing and deep breathing

CONTENT FOCUS

During the next several weeks, you'll be studying mammalian anatomy and physiology. To help you get a clear idea of the various organ systems and how they work, you'll be looking at the anatomy of the **fetal pig.**

Why fetal pigs? It may surprise you to learn that humans and pigs are very similar in anatomy. Today, pig organs and skin are frequently used for transplants and grafts to replace damaged human tissues. In addition, the skeleton of a fetal pig isn't fully calcified, making dissection easier to perform.

During extreme weather, farmers often sell surplus animals rather than run the risk of their dying from heat or cold. As part of the butchering process, all the organs, including the uterus, are removed. If fetal pigs are found in the uterus, they are preserved for educational purposes.

ACTIVITY 1   GETTING STARTED

1. Get the following supplies: **two long pieces of string, a tray, several paper towels, and some dissecting instruments (one scalpel, one pair of scissors, one large pair of forceps, and one blunt probe).**

2. Your laboratory instructor will distribute fetal pigs to each group or give you instructions for obtaining a pig.

3. Spread **two paper towels** on the tray. **Place the pig on its back on the tray.**

4. Tie the **ends** of the first string **tightly** around the two front hooves of your pig and slide it **underneath the tray.** Tie the **ends** of the other string around the two rear hooves and slide it underneath the tray.

5. Rotate the tray so that the pig's **tail is facing you.**

    Where is the **left side of the pig?** _____

Note:	

Keep the position of the pig in mind as you proceed with the dissection. The instructions refer to the pig's right and left sides, NOT your right and left sides.

ACTIVITY 2   FOLLOWING ANATOMICAL DIRECTIONS

Before we begin the dissection, it's necessary to understand some special terms that relate to anatomical locations and directions in the pig's body. Using **Figure 13-1** as a guide, use the correct anatomical terms to describe the location of the following:

1. The pig, as placed on the tray, is lying on its ___*back*___ surface.

2. The umbilical cord is located on the ___*top*___ side of the pig.

3. The attachment of the umbilical cord to the belly is ___*Yes*___ to the **hind** legs.

4. The attachment of the umbilical cord to the belly is ___*Yes*___ to the **front** legs.

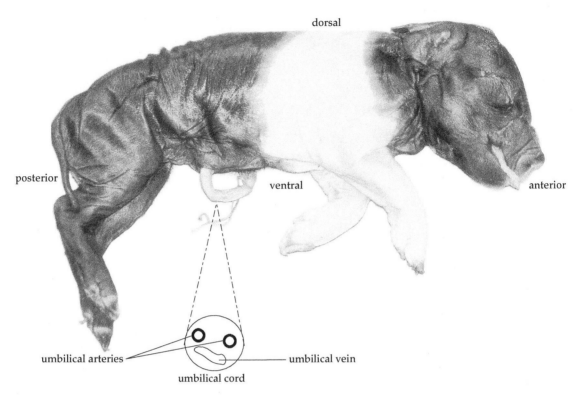

FIGURE 13-1. Anatomical Locations in the Pig

ACTIVITY 3   EXTERNAL ANATOMY OF THE FETAL PIG

1. Determine the **gender** of your pig, using **Figures 13-2 and 13-3** as a guide. There are two ways to tell males from females **externally.**

> **Note:**
>
> **Make sure you observe pigs of both genders.**

2. Look for the **anus,** which is located close to the base of the tail.

    Female pigs have a small, fingerlike projection, **just ventral to the anus.** This projection is called the **genital papilla.** On either side of the papilla, are the two **labia.** The **urogenital opening** is located between the two labia. The term **urogenital** refers to the double function of this opening. It allows for the excretion of urine (**"uro"** refers to the urinary system) and also leads to the **reproductive structures** (**"genital"** refers to the reproductive system).

> **Hint:**
>
> **If the papilla is not evident, you have a male pig.**

3. In **male pigs,** you'll notice the skin appears puffy or baggy just **anterior to the anus.** This is the **scrotum,** which contains the **testes.**

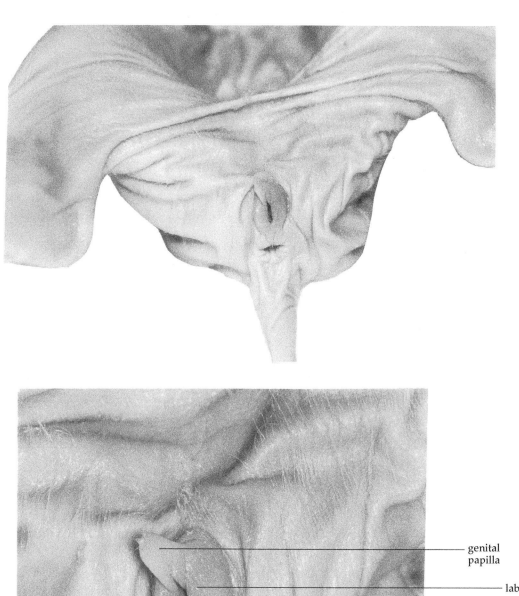

genital
papilla

labia

urogenital
opening

anus

FIGURE 13-2.   Female External Anatomy

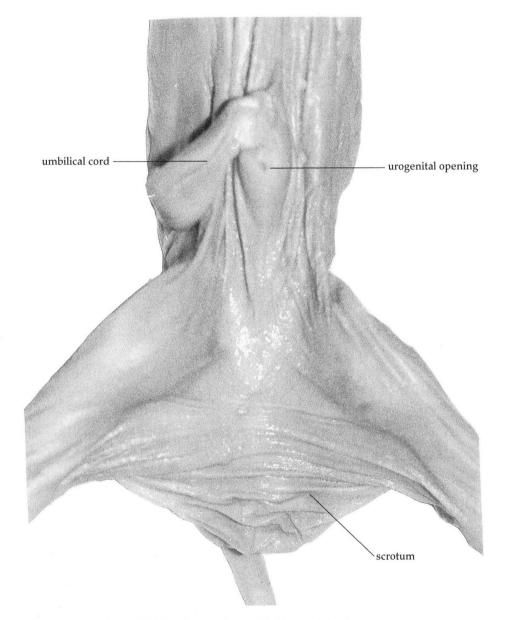

umbilical cord ———————— urogenital opening

scrotum

FIGURE 13-3. Male External Anatomy

4. The **urogenital opening** in the male is located just **posterior** to the attachment of the **umbilical cord** to the body wall. As with the female pigs, this opening functions in both the urinary and reproductive systems. Both urine and sperm are released here, but not at the same time.

    You can't see the **penis** on the exterior of the body because pigs, like many other four-legged animals, have a **retractable** penis, which can be seen when the animals reproduce. You'll notice that both male and female pigs have two rows of **nipples** on either side of the umbilical cord. As with humans, the nipples are only functional in females.

5. The body of the pig is divided into three regions: **head, neck,** and **trunk.** The trunk region, with the four legs and tail attached, is further subdivided into the **anterior** area, the **thorax** (chest area), and the **posterior** area, the **abdomen.**

    The thorax contains the heart and the lungs, enclosed by a protective rib cage. The abdomen contains the organs of the digestive system along with organs of many other important systems.

6. The **umbilical cord** attaches the fetus to the **placenta** in the mother.

    Using your scissors, make a cut across the umbilical cord to expose the two **umbilical arteries and the single, larger, umbilical vein** (see **Figure 13-1**).

    Push against the outside of the wall of the umbilical vein with a blunt probe. Repeat the process with an umbilical artery. Note that the wall of the umbilical vein is thinner and bends easily in comparison to the umbilical arteries.

7. Each pig has a **slit** cut in the side of the neck. This cut was made to inject the circulatory system with **colored latex** (a form of liquid rubber). The **arteries** have been injected with **pink latex** and the **veins** have been injected with **blue latex.** This will help you tell the blood vessels apart when you study the circulatory system.

ACTIVITY 4   DISSECTION OF THE MOUTH

1. Carefully, cut through **both sides of the jaw,** as illustrated in **Figure 13-4.** As you cut, **alternate** between the right and left sides until you can open the mouth wide.

Caution!
Young pigs have very sharp, pointed teeth. Be careful where you put your fingers while opening the mouth.

2. Look at the **roof** of the mouth and **feel its texture.** The rippled **anterior region** is hard because it's composed of **bone and cartilage.**

    This area is called the **hard palate.** Notice that the hard palate is divided into **two halves.** The line separating the left and right sections of the hard palate runs directly along the **body midline.**

    Label the **hard palate** and the **body midline** on the diagram in **Figure 13-5.**

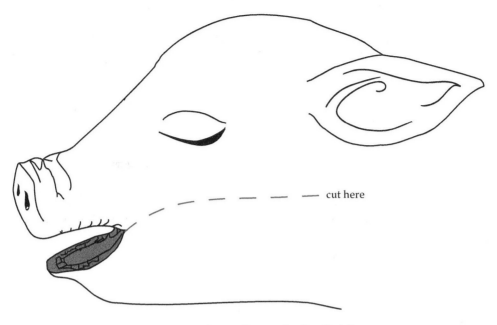

FIGURE 13-4.   Pattern for Jaw Incisions

3.   The **posterior region** of the roof of the mouth is the **soft palate.** Feel the texture of this area. **How is it different** from the hard palate area?

     It's hard

4.   The hard and the soft palates separate the mouth from the **nasal cavity.** As you may have experienced, drainage from the nose can enter the mouth, so there must be a **common area** where these two regions come together. This is the **pharynx.**

     The pharynx can be seen **posterior** to the soft palate.

     Insert the **tip of your blunt probe** into the opening of the pharynx and probe **anteriorly underneath the soft palate. Your probe is now in the nasal cavity.**

     Add labels for the **pharynx** and **soft palate** to **Figure 13-5.**

5.   Pull the mouth completely open and look down into the throat. There's a hood-shaped flap of tissue called the **epiglottis.** The epiglottis prevents food and liquid from entering the **glottis,** the opening to the air passageways (the prefix **"epi"** means *outside*).

     Add labels for the **epiglottis** and **glottis** to **Figure 13-5.**

6.   If you're eating and your food "goes down the wrong pipe," **which pipe** did the food enter? ___*glottis*___

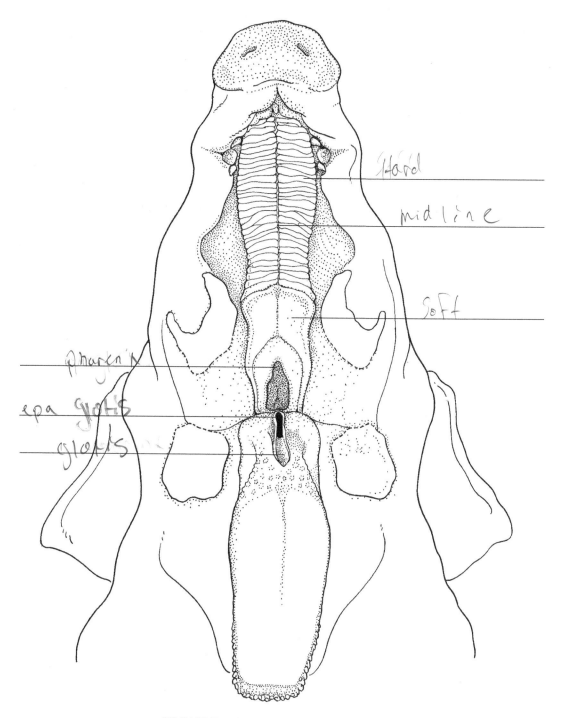

Hard

midline

Soft

Pharynx

epa glotis

glotis

FIGURE 13-5. Structures of the Mouth

Check your Figure 13-5 labels with your instructor before you continue.

ACTIVITY 5

DISSECTION OF THE NECK

Caution!
While cutting, be careful to keep the point of the scissors away from underlying organs.

Referring to **Figure 13-6**, make **only** the following cuts:

1. Cut the skin and muscle layers between the chin whiskers **(number 1)** and the anterior part of the sternum or breastbone **(number 2)** as in **Figure 13-6.**

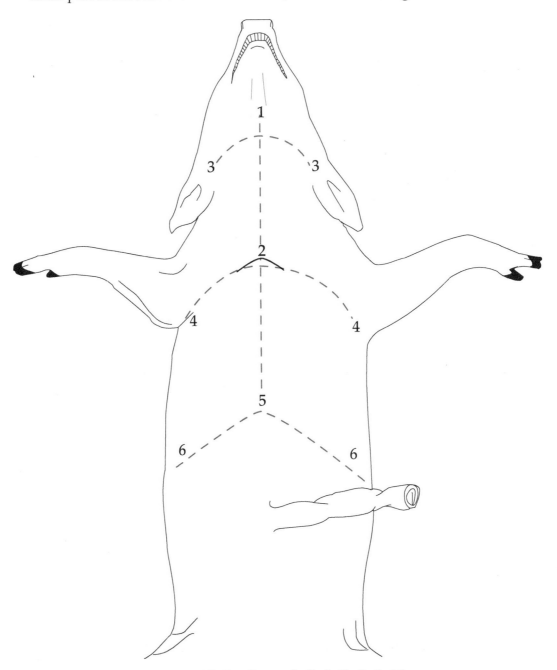

FIGURE 13-6. Pattern for Body Cavity Incisions

2. Make two additional incisions **across the throat (number 1 to number 3)** as shown in the figure.

3. Use your **blunt probe** to separate the muscles from the organs in the neck region. You'll expose the **thymus gland** on both sides of the neck. The thymus gland has an important role in **immune system function.**

4. With your fingers, feel around at the **anterior end** of the dissected area **(toward number 1 in Figure 13-6),** until you locate a hard bulblike structure. This is the **larynx.**

   Remove the muscle tissue that is covering the larynx until it can be clearly seen. The larynx is sometimes known as the **"voice box."**

   Name the **stringlike structures,** which **vibrate to produce sounds,** that are located inside the larynx. _____

5. Continue to expose the structures connected to the **posterior end of the larynx.** Remove the overlying tissues until you can see the tubular **trachea,** commonly known as the **windpipe.**

6. The trachea is supported along its length by many **cartilage rings.**

   What is the **function** of the trachea in your body?

   Why is the **presence of cartilage rings** in the trachea important for your survival?

7. At the **posterior end of the trachea,** you'll see a small, dark-brown organ, the **thyroid gland.** The thyroid gland produces **thyroid hormone,** which controls the body's **metabolic rate.**

8. Using your **blunt probe,** carefully separate the connective tissue from the sides of the trachea. The esophagus is **dorsal to the trachea.**

   Feel underneath the trachea and **lift the esophagus** with your probe. How is the esophagus **different from the trachea in structure?**

9. Is the esophagus part of the respiratory system? _____ **Explain** your answer.

Check your answers with your instructor before you continue.

10. On the side of the neck **opposite to the slit,** carefully remove the overlying muscle and connective tissue to expose the **carotid artery** and **two jugular veins.**

    These vessels will be running **parallel** to the trachea (see **Figure 13-7**).

11. The **carotid artery** is the **blood vessel closest to the trachea** and should be filled with **pink** latex. The **two jugular** veins should appear **blue.** The carotid artery carries blood rich in oxygen and nutrients toward the head. The jugular veins carry deoxygenated blood back toward the heart.

12. **Dissection Challenge!** Can you find the **vagus nerve** in the neck? It looks like a **flat,** white thread running close to the carotid artery and internal jugular vein. The vagus nerve is crucial for maintaining the normal state of homeostasis. Among other functions, this nerve **regulates heart rate, breathing,** and **digestive system activity.**

13. **Label** the following structures on **Figure 13-7: thymus, larynx, trachea, thyroid gland, carotid artery,** and **jugular veins.**

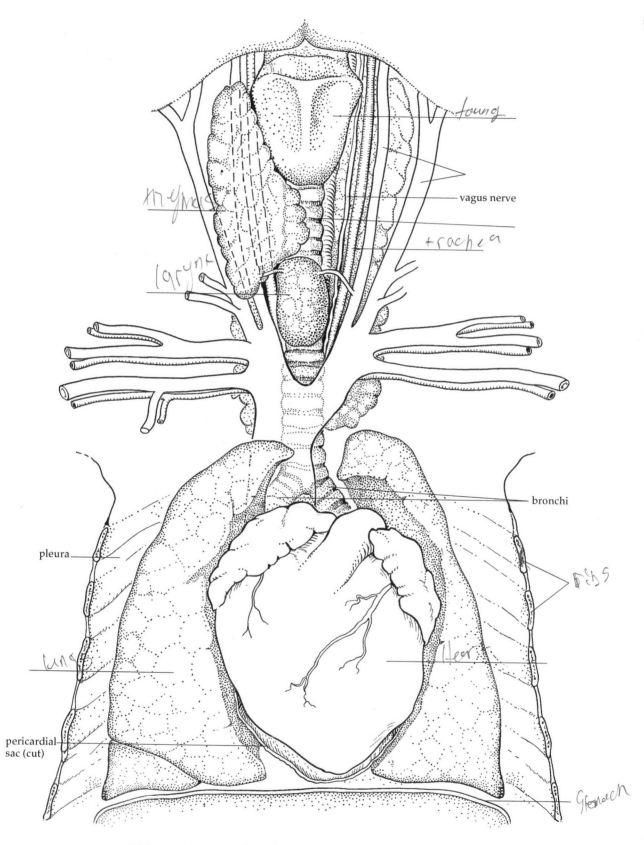

young

vagus nerve

thymus

trachea

larynx

bronchi

pleura

ribs

lungs

Heart

pericardial
sac (cut)

Stomach

FIGURE 13-7.   Organs of the Neck and Thoracic Cavity

✓ Comprehension Check

Fill in the blanks with the choice that is **most appropriate** for the listed functions. **Answers can be used only once.**

a. carotid artery   d. trachea   g. thymus
b. esophagus   e. thyroid gland   h. jugular vein
c. larynx   f. vagus nerve

1. _____ Tube that transports food to the stomach.

2. _____ A blood vessel carrying deoxygenated blood back to the heart.

3. _____ This structure produces a hormone that controls the body's metabolic rate.

4. _____ Tube held open by cartilage rings.

5. _____ A blood vessel carrying nutrient-rich blood to the head.

6. _____ This structure contains the vocal cords.

7. _____ Plays an important role in immune system defense.

8. _____ Regulates heart rate, breathing, and digestive system activities.

ACTIVITY 6   DISSECTION OF THE THORACIC CAVITY

1. With your fingers, find the **anterior end of the sternum.** With your **scissors,** cut through **the sternum and the rib cage** until you reach the end of the rib cage (cut from **number 2 to number 5** and from **number 2 to number 4** on **Figure 13-6**). This will give you a clearer view of the ribs and help you pinpoint the site of your next incision.

2. **Make two more incisions** toward the sides of the body to completely open up the chest cavity **(number 5 to number 6).** As you cut, you'll be separating the **ribs** from the **diaphragm,** a large muscle that extends **completely across** the body cavity. **The diaphragm expands the chest cavity so that air flows into the lungs.**

3. The thoracic cavity is lined with layers of smooth membranes called the **pleura.**

   As you look into the thoracic cavity, the **two lungs** are easy to see. Each lung is enclosed in a **pleural cavity,** lined by the **pleural membranes.**

   The lungs are composed of millions of tiny sacs, which are the sites of gas exchange for the whole body. **Oxygen enters the lungs and carbon dioxide is removed.**

> ### Note:
> The trachea extends posteriorly from the larynx and divides into branches called bronchi, which extend to the lungs. The branches can't be seen at this stage of the dissection, but we'll examine them later.

4. Located between the two lungs is the **heart.** The heart is a muscular organ that **pumps blood to the lungs and the body tissues.** The heart of the pig is located in the center of the chest cavity. This is also true of the human heart (not on the left side as is commonly believed).

    The **pericardium** (also called the **pericardial sac**) is a tough membranous sheet that completely encloses the heart. **Remove the pericardial sac** so that you can get a better look at the heart.

5. **Add the following labels** to the diagram in **Figure 13-7: lungs, heart, ribs, and diaphragm.**

Check your Figure 13-7 labels and your answers with your instructor before you continue.

6. After you complete your dissection:

    ■ Get a **plastic storage bag** for your pig. Place a strip of tape near the bottom of the bag and write your group's names on the tape.

    ■ Slide the pig off the tray with the **string still attached** to the limbs.

    ■ Wrap the pig in **damp paper towels** and seal it in the storage bag. Your instructor will tell you where to store your pig.

    ■ **Wash and dry your trays and tools. Your instructor will give you specific instructions for disposing of any remaining pig tissues.**

ACTIVITY 7   MEASURING LUNG CAPACITY

Just looking at the lungs doesn't give you a very good idea of how lungs function. In the following activities you'll experiment on yourself and your lab partners and develop some ideas about how respiratory function adjusts to your body's needs when you exercise.

The amount of air that enters and leaves your lungs each time you take a breath is called the **tidal volume.** In most men and women, the tidal volume is about **500 cc (half a liter).**

You can increase the amount of air inhaled and exhaled by deep breathing. The **maximum** amount of air that you can move in and out during a single breath is called the **vital capacity.**

1. Work **with a partner. Each** student will perform this experiment and record his/her **individual** results.

Caution!
DON'T use the balloon method if you have respiratory or heart problems!

2. Get **a metric ruler, a clamp, and one balloon for each person.**

Note:
Read through the directions COMPLETELY before beginning!

3. Measure your respiratory volume according to the following directions.

   a. Sit down and relax.

   b. **Inhale normally** and then **exhale** only that normal breath into the balloon.

Note:
If you have trouble inflating the balloon, stretch it several times and try again.

   c. **Immediately** twist the balloon several times and clamp it shut so that no air escapes.

   d. Place the tip of a pencil vertically onto the "zero" cm mark. Place the balloon on its **side** next to the pencil (**position A in Figure 13-8**).

Holding the balloon in position, move your pencil point from the zero mark and place it on the ruler at the right edge of the balloon (**position B in Figure 13-8**).

Record the diameter of the balloon in **Table 13-1.**

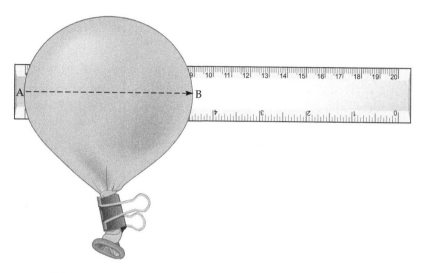

FIGURE 13-8.   Method for Measuring Balloon Diameter

4.   Repeat the entire process **twice more. Record** the results in **Table 13-1.**

TABLE 13-1	
MEASURING LUNG CAPACITY	
DIAMETER (cm) OF BALLOON FILLED WITH NORMAL BREATH:	
Trial 1	
Trial 2	
Trial 3	
Average	
Volume of air (cc) in balloon filled with normal breath:	
DIAMETER (cm) OF BALLOON FILLED WITH DEEP BREATH	
Trial 1	
Trial 2	
Trial 3	
Average	
Volume of air (cc) in balloon filled with deep breath:	

5. Calculate the **average balloon diameter** for your three trials of exhaled air **at rest.** **Record** the average in **Table 13-1.**

6. a. **Inhale as much air as you can** (with **only one** very deep breath) **and exhale as much air as you can** into the balloon. **Clamp and measure** the balloon as in **step 3** above.

   b. Record your results in **Table 13-1.**

7. Calculate the **average balloon diameter** for your three trials of exhaled air **while deep breathing. Record** the average in **Table 13-1.**

8. **Use the graph** in **Figure 13-9** to **convert** the average diameter of the balloon filled by **a normal breath** and the balloon filled by **a deep breath** into **cubic centimeters (cc) of lung volume.**

   Record the information in **Table 13-1.**

9. What is your **tidal volume?** _____ cc

   What is your **vital capacity?** _____ cc

Note:
Your measurement of tidal volume may be slightly higher than your actual tidal volume, since you also exhaled the reserve air held in your lungs.

---

10. Which volume of air is exchanged while you're sleeping? _____

    Which volume of air is exchanged while you're running? _____

    Explain your answer.

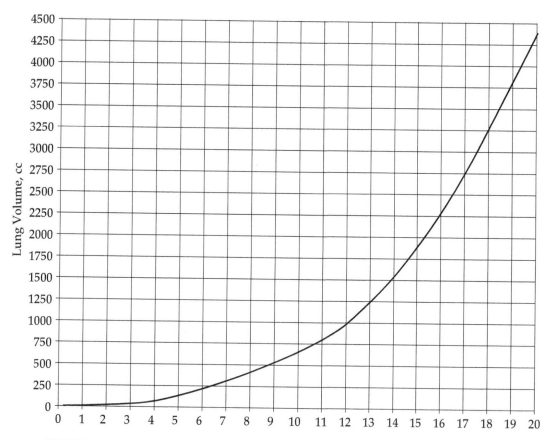

FIGURE 13-9. Relationship Between Balloon Diameter and Volume of Exhaled Air

Check your answers with your instructor before you continue.

Self Test

Match the definitions in the right column with the terms in the left column. **Each answer can be used only once.**

1. _____ anterior   a. back

2. _____ dorsal   b. tail end of the body

3. _____ posterior   c. head end of the body

4. _____ ventral   d   belly side

5. Your lab partner arrived late to class on the day you started your fetal pig dissection. You've already selected a male pig for your dissection and you want your partner to choose a female. **Describe** how your partner can recognize and select a female pig from the container. **Be specific.**

6. Erin is playing shortstop on her college baseball team when she is hit in the throat by a line drive. On the way to the hospital, she has a very difficult time breathing. Which of the following is the most likely cause of her problem? **Explain** your answer.

   a. bruised diaphragm   d. crushed trachea
   b. concussion   e. blocked esophagus
   c. broken rib cage

7. Name **one part of the digestive system** that's located in the **thoracic** cavity.

8. Name a structure that lies **between** the thoracic and abdominal cavities.

9. When breathing, you're never able to completely fill your lungs with fresh air. A certain volume of stale air (with the oxygen removed) always remains in the air passageways. In some lung diseases, such as **emphysema,** large amounts of stale air accumulate and can't be expelled. The lungs are filled with air that is useless for gas exchange. **Vital capacity** is significantly less than in normal lungs.

    Predict what will happen when a person with emphysema exercises heavily. **Explain** your answer.

10. In regard to your answer to question 9, explain how lack of air affects the body tissues. In your answer, use the following terms: **energy, oxygen, mitochondria, ATP, and red blood cells.**

14

Organ Systems in the Abdominal Cavity

Objectives

After completing this exercise, you should be able to:

- identify and explain the function of each of the major structures of the abdominal cavity
- explain how the stomach and small intestine are specialized to perform specific functions
- compare and contrast anatomical features of the pig with those of humans
- apply your knowledge of the benefits of increasing surface area in the radish root to similar modifications in the intestine and lung
- relate your observations of earthworm locomotion to peristaltic action in organs of the digestive tract

ACTIVITY 1   DISSECTION OF THE ABDOMINAL CAVITY

1. **Get your pig** from the storage container and set it up on a tray as you did in your previous pig dissection. Get a set of **dissecting instruments.**

2. Use your **scissors** to cut through the **skin and muscles** in the abdominal region.

    Pull up on the umbilical cord to hold the skin away from the abdominal organs and cut from **number 5 to number 7** (see **Figure 14-1**).

    Make two more incisions **around the umbilical cord (number 7 to number 8).**

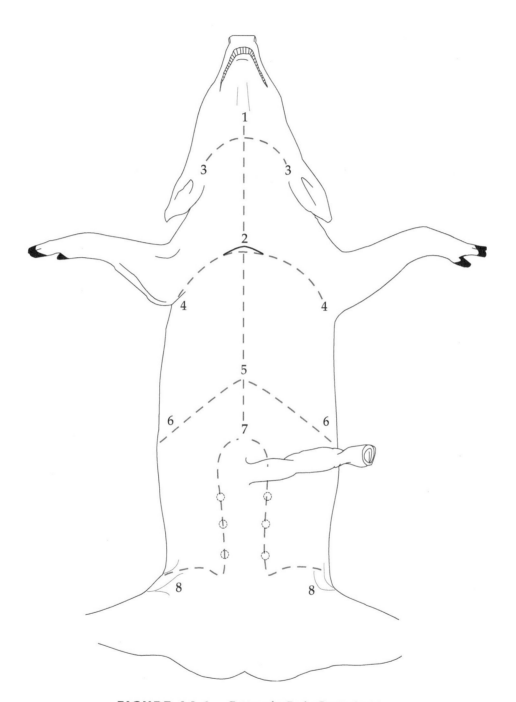

FIGURE 14-1. Pattern for Body Cavity Incisions

3. The flap of tissue created by these cuts contains the **umbilical vein, the two umbilical arteries,** and the **urinary bladder.**

   Cut the umbilical vein **close to the umbilical cord.** This will allow you to turn back the flap so that it lies between the hind legs.

Hint:

If needed, rinse out the entire body cavity and blot it dry with paper towels. This will make it easier to locate and identify the abdominal organs.

---

4. Just as in the thoracic cavity, the abdominal cavity is sealed with a smooth membrane called the **peritoneum.**

5. The largest organ in the abdominal cavity is the **liver.** It's reddish-brown with several lobes and is located just **posterior to the diaphragm.**

   The liver has many functions including **storage of energy reserves, detoxification of poisons, and bile formation.**

6. And speaking of bile, lift the **right lobe** of the liver and locate the **gall bladder,** a small saclike organ.

   The gall bladder **stores bile produced by the liver. Bile aids in fat digestion.** Bile is released through the **bile duct** into the small intestine, where fat digestion takes place.

7. Raise the liver to locate the **stomach,** a large, hollow bag located on the left side of the body.

   The stomach stores food and releases it gradually into the small intestine. Digestion of proteins begins here.

8. The long, dark red organ lying to the left of the stomach is the **spleen.**

   The spleen **functions as part of the immune system** and also **removes damaged and worn-out blood cells from circulation.**

   Within the abdominal cavity, the organs are held in position by membranes called **mesenteries.** You can see this membrane attaching the spleen to the stomach.

9. The **pancreas** is located between the stomach and the small intestine. To expose this organ, lift the stomach and use your **blunt probe** to dissect through the membranes in this area. The pancreas is an **elongated, pale, granular** organ positioned **across the body cavity.**

    The pancreas **secretes digestive enzymes** through a duct into the small intestine. It also functions as an **endocrine organ,** producing **two hormones that control blood sugar level.**

10. The **small intestine,** on the right side of the abdominal cavity, is a long, thin tube. Most chemical digestion occurs here. In the small intestine, **enzymes break down proteins, fats, carbohydrates, and nucleic acids.**

    Hold up one of the loops of the small intestine and notice that it is held together by **mesenteries.** If you look closely, you can see **arteries and veins** fanning out across the membrane.

    Mesenteries provide a **support structure** for blood vessels and nerves leading to all the abdominal organs.

11. The **large intestine** is on the left side of the abdominal cavity. The main function of the large intestine is the **reabsorption of water from wastes,** preventing dehydration. The concentrated wastes are stored in the **rectum** and eliminated through the **anus.**

12. Lift the intestines on either side of the body and you'll see a **kidney.** The kidneys **remove nitrogen-containing wastes and form urine.**

13. The **urinary bladder,** located between the two umbilical arteries, is a **temporary storage organ for urine.** When the bladder fills, nerve endings in its muscular walls are stimulated, causing the muscles to contract for urination.

14. **Label** the following structures on **Figure 14-2:** liver, gall bladder, stomach, spleen, pancreas, small intestine, large intestine, mesenteries, kidney, and rectum.

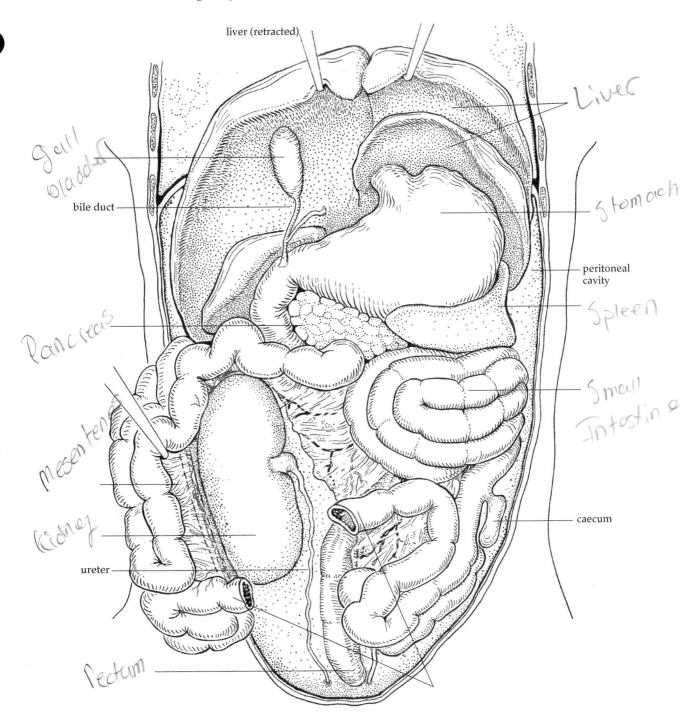

liver (retracted)

Liver

Gall bladder

bile duct

Stomach

peritoneal cavity

Spleen

Pancreas

Small Intestine

Mesentery

Kidney

caecum

ureter

Rectum

FIGURE 14-2. Organs of the Abdominal Cavity

Check your figure labels with your instructor before you continue.

ACTIVITY 2   THE URINARY SYSTEM

Waste products are removed from the body by several systems. For example, **carbon dioxide,** a waste product of **cell respiration,** is removed by the **lungs. The nitrogen-containing** waste products of **protein digestion** are excreted by the urinary system.

The urinary system includes the two **kidneys,** which **produce urine;** the two **ureters,** which **carry the urine to the bladder;** the **bladder,** which **stores the urine;** and the **urethra,** which **transports urine** out of the body through the **urogenital or urinary opening** (depending on your sex).

The kidneys also have an important function in maintaining **homeostasis** (keeping internal body conditions stable). The body must maintain adequate levels of salts, sugars, and other substances in the bloodstream, while disposing of the excess. The kidneys play a key role in this process, keeping **pH** values, **fluid and salt content,** and **plasma levels of valuable materials** nearly constant.

1. Push the intestinal mass to the **left side** of the abdominal cavity, exposing the **right kidney.** The kidneys are located along the **dorsal wall of the abdominal cavity** (see **Figure 14-2**).

2. You'll notice that the kidney is covered by a thick layer of connective tissue, the **peritoneum.**

    Using the blunt probe, carefully remove the peritoneum, uncovering the right kidney and ureter.

    In addition to the ureter, you'll see the **renal artery and vein.** The renal **artery** carries blood **to the kidney for filtration.** The blood **returns** to the body circulation through the **renal vein.**

3. Each kidney is drained by a **ureter,** a tube leading from the kidney to the urinary bladder.

    Locate the ureter and follow it to the **bladder,** which is located in the flap of tissue created by your **incisions around the umbilical cord in Activity 1.** The bladder can be found **between the two umbilical arteries.**

    The **posterior end of the bladder** narrows to form a tube, the **urethra,** which transports urine out of the body. Most of the urethra can't be seen, because it's positioned deeper in the pelvic cavity.

4. Label the following structures on **Figure 14-3: kidney, urethra, and bladder.**

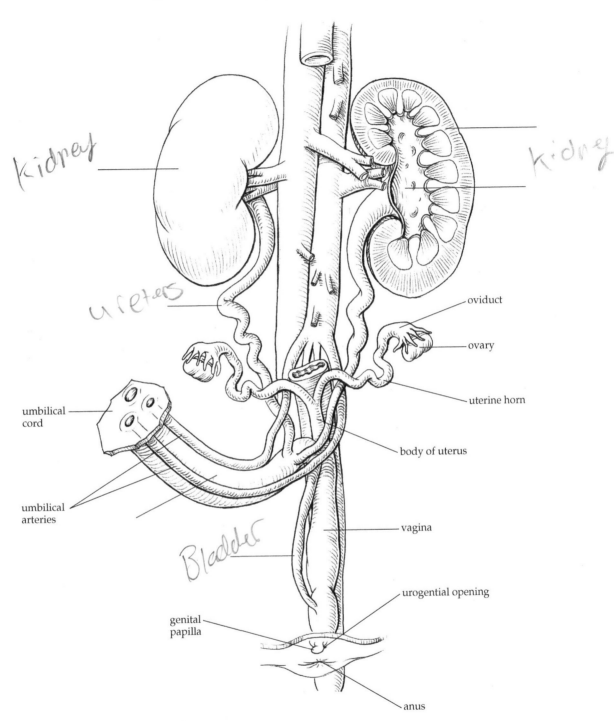

kidney

kidney

ureters

Bladder

oviduct

ovary

uterine horn

umbilical
cord

body of uterus

umbilical
arteries

vagina

urogential opening

genital
papilla

anus

FIGURE 14-3. Urinary System of the Female Pig

5. Cut the **ureter** and **renal blood vessels.** Remove the **intact right kidney** from the abdominal cavity. If necessary, remove additional peritoneum so that the kidney can be **removed without damage.**

6. Make an incision through the long axis of the kidney, cutting it into two equal halves (like opening a book).

   Within the kidney, you'll see an **outer area** of densely packed tissue, the **cortex.** **Urine formation begins** in microscopic structures within this region. The **inner** section, the **medulla,** appears fibrous. This area **collects the urine** and funnels it to the ureter.

   Observe the connection of the ureter to the kidney near the center of this region.

7. Label the following structures on the dissected kidney in **Figure 14-3: cortex, medulla, and ureter.**

✔ Comprehension Check

Fill in the blanks with the choice that is most appropriate to describe the function of each part of the urinary system. **Answers can be used more than once. Some questions have more than one correct answer.**

a. ureter   d. renal artery   g. urinary bladder
b. kidney   e. renal vein   h. medulla
c. urethra   f. peritoneum   i. cortex

1. _____ This membrane holds the kidneys in position.

2. _____ If a drop of urine is in the ureter, which **TWO** structures will it pass through next **(in sequence)?**

3. _____ Carries blood high in metabolic wastes to the kidney.

4. _____ Returns filtered blood to body circulation.

5. _____ When this structure is full, nerve endings signal the urge to urinate.

6. _____ Urine formation begins in this region of the kidney.

7. _____ Either urine or semen could exit the body through this tube.

8. _____ Plays a vital role in maintaining homeostasis in the body.

9. _____ This portion of the kidney funnels urine to the ureter.

Check your answers and Figure 14-3 labels with your instructor before you continue.

ACTIVITY 3   INTERNAL STRUCTURE OF THE STOMACH

1. With your fingers, locate the **pyloric sphincter** of the fetal pig. It'll feel like a **small, hard mass** in the stomach wall. Using the same method, locate the **cardiac (gastroesophageal) sphincter.**

    Referring to **Figure 14-4,** use your **scissors** to remove the entire stomach, being careful to include **both sphincters.**

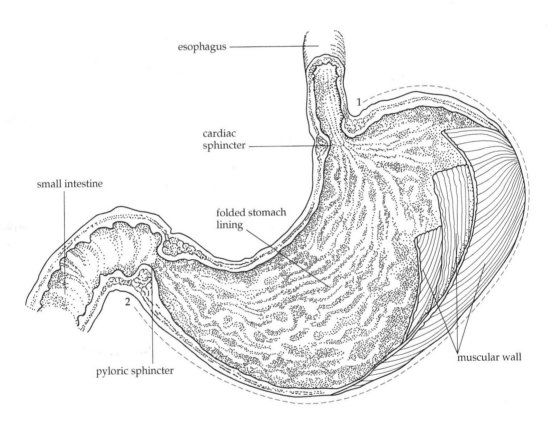

FIGURE 14-4.   Internal Structure of the Stomach

2. Following the **dotted line** in **Figure 14-4,** cut from **number 1 to number 2** and open up the stomach.

    Rinse the inside of the stomach and **blot it dry** with paper towels.

3. Examine the **interior** of the stomach. **Observe** that the stomach wall is folded, like an accordion.

   Imagine that, disregarding what you know about dietary guidelines, you stuff yourself with two cheeseburgers, a super-size order of fries, an apple pie, and a chocolate shake.

   During this intake, what happens to the **size** of your stomach? _Increases_

   How are the **folds in the stomach wall** related to this change in size?

   expansion

4. A sphincter is a **ring of muscle** that surrounds the opening of a hollow organ. When the sphincter muscles **contract,** the opening is shut.

   Visualize what happens when you pull the ties on a plastic trash bag and the bag closes. This is similar to the action of a sphincter muscle.

   The stomach has **two sphincters.** One is between the esophagus and the stomach (the **cardiac sphincter**) and the other is between the stomach and the small intestine (the **pyloric sphincter**).

   (Circle one answer.)

   If there's food in the esophagus, the cardiac sphincter will be **opened** / closed.

   What happens to the food in the **stomach** while **both sphincters are closed?**

   Digestion

   If the **pyloric sphincter is open,** where does the food go? _Small intestine_

5. With one hand, feel the pyloric sphincter. At the **same time,** use the other hand to feel the cardiac sphincter.

   How are the two sphincters **different?**

   Cardiac Sphincter is larger than the Pyloric Sphincter

   Which sphincter do you think is **stronger?** *Cardiac*

   The **cardiac sphincter** prevents food in the stomach from backing up into the esophagus.

   The **pyloric sphincter** controls the **rate** at which food enters the small intestine. No matter how full your stomach is, food is always released **gradually** into the small intestine.

✓ Comprehension Check

1. Considering your examination of the stomach, **explain** what happens during **vomiting.**

2. What would happen if your **pyloric sphincter** weren't able to **close?**

3. What would happen if your **pyloric sphincter** weren't able to **open?**

4. What would happen if the **cardiac sphincter** didn't close completely and stomach juices were allowed to enter the esophagus?

Check your answers with your instructor before you continue.

ACTIVITY 4   EXAMINING THE SMALL INTESTINE

1. Hold up a loop of the small intestine and **examine the mesenteries.**

    In addition to holding the small intestine in position, the mesenteries provide a passageway for a **large number of blood vessels.**

2. Using your **scissors,** carefully **snip through the mesenteries** and start separating the loops of the small intestine.

    Remove the small intestine.

3. Lay the intestine out on the laboratory counter and use a **meter stick** or **yardstick** to measure the length.

    The length of my pig's small intestine is _____.

    (Circle one answer.) The small intestine is **shorter / longer** than I expected.

    If the small intestine is longer than you expected, you might be interested in the origin of the name "small" intestine. The intestine is named for its **small diameter (and not its length).**

    The large intestine has twice the diameter of the small intestine **(but only half the length).**

4. Although each person is slightly different, the small intestine in humans averages about **20 feet (6 meters)** in length.

    Using your knowledge of the digestive system, **explain** how increasing the length of the small intestine would help this organ **perform its functions.**

5. Remove approximately **two inches (5 cm)** from the small intestine. Using your **scissors,** cut the piece of intestine **open lengthwise** and place it on the stage of the **dissecting microscope.**

6. Observe that the inside of the small intestine is **not smooth.**

   The interior wall is **folded,** and **each fold is carpeted** with slender fingerlike projections called **villi** (singular is **villus**).

7. Although you can't see them at this magnification, each of the villi is covered, in turn, with even more tiny projections called **microvilli** (see **Figure 14-5**).

   Located inside each of the villi is a network of capillaries and lymphatic vessels **(lacteals).** These vessels pick up nutrients as they are absorbed, so that they can be distributed to all parts of the body.

   The small intestine is quite a narrow tube. However, if the villi and the microvilli in the intestinal tract were laid out flat, the surface area would be **half the size of a basketball court!**

8. After you complete your dissection:

   ■ Get a **plastic storage bag for your pig. Place a strip of tape near the bottom of the bag and write your group's names on the tape.**

   ■ Slide the pig off the tray with the **string still attached** to the limbs.

   ■ Wrap the pig in **damp paper towels** and seal it in the storage bag. Your instructor will tell you where to store your pig.

   ■ **Wash and dry your trays and tools. Your instructor will give you specific instructions for disposing of any remaining pig tissues.**

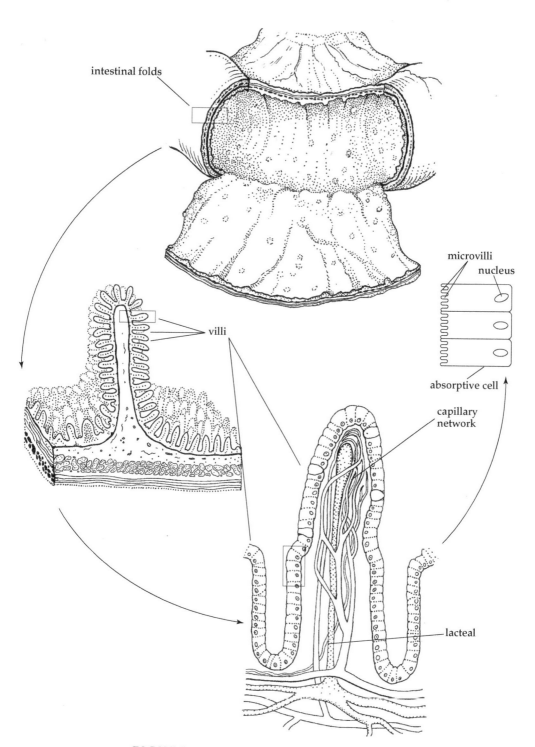

FIGURE 14-5. Interior of the Small Intestine

ACTIVITY 5   THE IMPORTANCE OF SURFACE AREA

1. **Why does the small intestine need such a large surface area?** The need to increase surface area is common throughout the plant and animal kingdoms. On the laboratory counter, you'll find a demonstration of one method used to increase surface area in a plant root **(a radish seedling).**

    (Circle one answer.) The radish root tip appears **smooth / fuzzy.**

    What **two** important substances enter a plant through the **roots?**

    _____   _____

2. The many slender extensions you can see covering the root of the radish seedling are called **root hairs.**

    Use **Figure 14-6** to determine how **root hairs** help plant roots function.

    In **both** the pictures in **Figure 14-6, the dark line represents the outside of the root surface.**

    Following the dark line, **place an X in every square that the line passes through** (any square that touches the exterior of the root surface).

    Place only one X in each square, **even if the line passes through a square twice.**

    A few Xs have been drawn into the figure to get you started.

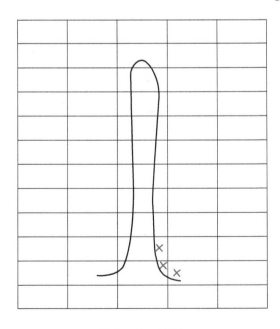
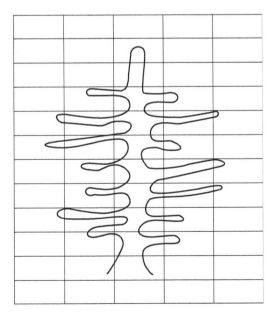

Without Root Hairs   With Root Hairs

FIGURE 14-6.   Comparison of Surface Area With and Without Root Hairs

3. For each picture in **Figure 14-6, count the number of squares with an X** and enter your results below:

   Without root hairs _____   **With root hairs** _____

4. Imagine that each graph square represents **one drop of water.** Which root (with or without root hairs) can absorb the **greatest number of water droplets at one time? Explain** your answer.

✓ Comprehension Check

1. Imagine that each graph square represents **one nutrient molecule.** Which type of intestinal lining (with or without villi) can absorb the **greatest number of nutrient molecules at one time? Explain** your answer.

2. Considering your answer to the above question, explain how **microvilli** are beneficial for small intestine function?

Check your answers with your instructor before you continue.

ACTIVITY 6
A CLOSER LOOK AT THE INSIDE OF THE SMALL INTESTINE

1. Referring to **Figure 14-7,** locate the **villi** (projecting toward the center of the tube).

2. At the base of the villi, you'll see many **small,** circular structures.

    These are the **intestinal glands.** Intestinal glands produce **digestive enzymes** that perform the final breakdown of food molecules.

    Add a **label** for **intestinal glands** to the diagram in **Figure 14-7.**

3. Opposite the intestinal glands, you'll see a cluster of **larger** circles.

    These are **lymphoid nodules,** small lymph nodes that act as **bacterial filters** in the digestive, respiratory, and urinary tracts.

    Which other body locations contain **lymph nodes?**

    Add a **label** for **lymphoid nodules** to the diagram in **Figure 14-7.**

4. You can see **two layers of muscle tissue** around the outside of the intestine.

    The **innermost** layer of muscle fibers runs in a circle around the tube and is referred to as **circular muscle.**

    The **outermost** layer has the muscle fibers arranged along the **length** of the tube and is called **longitudinal muscle.**

    Add **labels** for the **circular and longitudinal muscle** layers to the diagram in **Figure 14-7.**

5. The **circular and longitudinal** muscle layers **alternately** contract, producing mixing movements and moving food through the small intestine with waves of **peristalsis.**

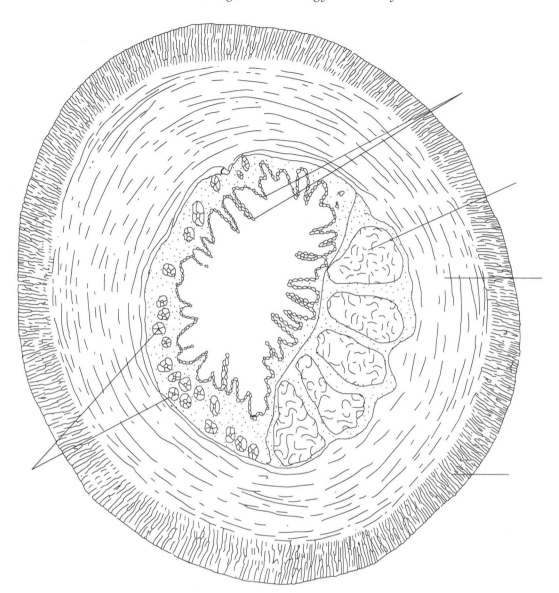

FIGURE 14-7. Cross Section of the Small Intestine

Check your figure labels with your instructor before you continue.

ACTIVITY 7   EARTHWORM LOCOMOTION

To see waves of **peristaltic action,** we'll examine locomotion in an earthworm.

Earthworm movement is a result of alternating muscle layers in the body wall. Earthworms move forward by alternating contractions of the **circular and longitudinal muscle** layers in the body wall. Segments may also move in the reverse direction, allowing the earthworm to move backward.

1. Work in groups. **Get an earthworm and several clean, dampened paper towels.**

Note:
Wash your hands before touching the worms!

2. **Lay out** the dampened paper towels on your laboratory counter and **place the earthworm** on the towels.

    Observe the **muscle contractions** as the earthworm moves forward. From your observations, **describe how the earthworm's body changes** as it moves.

3. For a closer look at the alternating action of circular and longitudinal muscles, place the earthworm on the stage of your **dissecting microscope.** Similar muscle contractions in the small intestine mix and move food.

    When you've completed your observations, **return the earthworm** to the container on the supply table.

4. How is the **process of peristalsis** related to the earthworm's locomotion? (If you need more information to answer this question, refer to your textbook.)

5. **(Circle one answer.)** The circular and longitudinal muscles in the small intestine are composed of **smooth muscle tissue / skeletal muscle tissue / cardiac muscle tissue.**

Self Test

Fill in the blanks with the **most appropriate** answer. **Answers can be used only once.**

a. lymphoid nodules
b. root hairs
c. villi
d. circular muscle

e. longitudinal muscle
f. peristalsis
g. surface area
h. mesenteries

1. _____ Membranes that hold the small intestines in position.

2. _____ Small projections that increase surface area in the small intestine.

3. _____ Involuntary muscle running **around** the intestine.

4. _____ Contractions that move food through the digestive tract.

5. _____ The **outermost** muscle layer in the small intestine.

6. _____ Has a **function similar** to the villi of the small intestine.

7. Which of the following organ systems are represented in the **abdominal cavity?** **Circle ALL correct answers** and **give an example of an abdominal organ or structure** that belongs to each system.

ORGAN SYSTEM	EXAMPLES
a. Digestive system	
b. Respiratory system	
c. Urinary system	
d. Circulatory system	

8. When you experience **heartburn,** which **stomach** structure isn't functioning correctly? **Explain** your answer.

9. **Bulimia** is an eating disorder that involves forced vomiting. Over a period of years, bulimics gradually lose the enamel coating from their teeth. Using your knowledge of the digestive system, **explain** what causes the loss of tooth enamel.

10. **Urinary incontinence** (inability to control urine flow) frequently occurs in the elderly. Which organ of the urinary system may be malfunctioning? Which structure is malfunctioning in the organ you named? **Explain** your answer.

11. **Figure 14-8** contains two designs for air sacs in a human lung. Which will be the **most efficient** in taking in oxygen and removing carbon dioxide? **Explain** your answer.

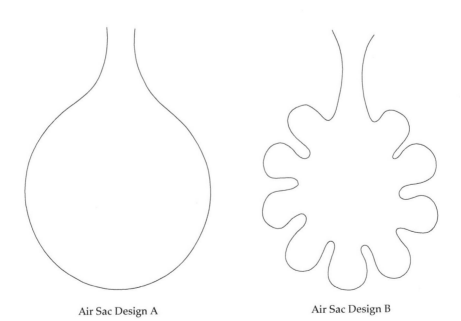

Air Sac Design A   Air Sac Design B

FIGURE 14-8. Comparison of Air Sac Designs

Complete the crossword puzzle in **Figure 14-9** to review the structures of the pig and their functions.

ACROSS

1. Pair of blood vessels that carry oxygen-poor blood that is high in wastes from the fetus to the placenta; called the umbilical _____
2. Most chemical digestion occurs here; enzymes break down proteins, fats, carbohydrates, and nucleic acids
3. External genital structure in female pigs
4. Part of the immune system; site of T-cell development
5. Muscular tube that connects the mouth to the stomach
6. Part of the immune system; blood-filtering organ and storage site for red blood cells and some white blood cells
7. Thin connective tissue membranes that hold the internal organs in place and provide a passageway for blood vessels and nerves that supply the organs
8. Bony roof of the mouth
9. Houses the vocal cords, the vibrations of which make speech possible; composed of cartilage
10. Fleshy sac that contains the testes in males
11. Endocrine (hormone-producing) gland; produces a hormone that helps in regulating the metabolic activity of the body
12. Secretes digestive enzymes into the small intestine; also functions as an endocrine organ, producing two hormones that control blood sugar levels
13. Has many functions, including storage of energy reserves, detoxification of poisons, and bile formation

DOWN

1. Stores bile formed by the liver; releases it through the bile duct into the small intestine (for fat digestion)
2. Baglike organ that stores food and releases it gradually into the small intestine; mechanical and some chemical digestion occurs here
3. Air channel to the lungs; surrounded by tough, elastic rings of cartilage
4. Excretory organ that stores urine
5. Hood-shaped flap of tissue that covers the opening to the trachea; prevents food and liquid from entering the air passageways
6. Composed of millions of tiny sacs, which are the site of gas exchange for the whole body; oxygen enters and carbon dioxide is removed
7. Paired blood vessels that drain deoxygenated blood from the head region back to the heart; called the jugular _____
8. Excretory organ that removes nitrogen-containing wastes and maintains the proper solute concentration of body fluids
9. Digestive organ whose main function is the reabsorption of water from wastes; prevents dehydration
10. Muscular pump consisting of four chambers; pumps blood to the lungs and the body tissues

FIGURE 14-9. Crossword Puzzle

The Reproductive System

Objectives

After completing this exercise, you should be able to:

- identify and explain the function of each of the major organs that make up the male reproductive system
- identify and explain the function of each of the major organs that make up the female reproductive system
- compare and contrast fetal pig reproductive structures with the comparable structures in humans
- trace the path of gametes through the male and female reproductive systems
- correlate sperm production with meiotic divisions
- identify the stages of follicular development within the ovary
- correlate the stages of follicular development to hormone production
- trace the stages of egg and follicle development through ovulation and corpus luteum formation

CONTENT FOCUS

Understanding the reproductive system is important when considering the many issues of reproductive health that face today's citizens. Issues such as endometriosis, sexually transmitted infections, and control of fertility are all better understood when you have a firm knowledge of reproductive anatomy and physiology.

The urinary and reproductive systems of mammals contain several shared structures, so, as part of your preparation for this dissection, you may wish to review the urinary structures (and their functions) that you examined in Exercise 14.

ACTIVITY 1

DISSECTION OF THE PIG REPRODUCTIVE SYSTEM

Although you're looking at the reproductive organs of male and female fetal pigs, most of the structures and their functions are similar to those found in the human reproductive system.

1. Get your pig from the storage container and set it up on a tray as you did in previous dissections.

   Get a **set of dissecting instruments.**

2. **Based on the gender of your pig,** follow the appropriate instructions below.

   When you've completed your dissection, examine the organs and structures in a pig of the opposite gender.

Dissection of a Male Fetal Pig

1. Locate the urogenital opening, posterior to the umbilical cord. Pinch the skin around the opening. You should feel a muscular tube — the **penis.** You cannot see the **penis** on the exterior of the body since pigs, like many other four-legged animals, have a **retractable** penis.

2. Beginning at the urogenital opening, use your scissors to cut along the body midline **through the skin only** for a couple of inches.

   Using a blunt probe, locate the **urethra,** a white tube beneath the skin. The urethra **transports urine** out of the body through the **urogenital opening.** When the penis is inserted into the vagina, sperm pass through the urethra into the female reproductive tract.

   As you continue your dissection, refer to **Figure 15-1** to see the positions of the various organs.

   Continue your cut through the skin in a posterior direction to expose the entire length of the urethra. As you cut, **be careful not to disturb the testes,** which can be seen on either side of the urethra in the scrotal sac.

3. Using scissors, remove the **tissue between** the exposed urethra and the **urinary bladder** (which is located between the two umbilical arteries), but do not remove the bladder.

   The **urinary bladder** is a **temporary storage organ for urine.** When the bladder fills, nerve endings in its muscular walls are stimulated, causing the muscles to contract for urination.

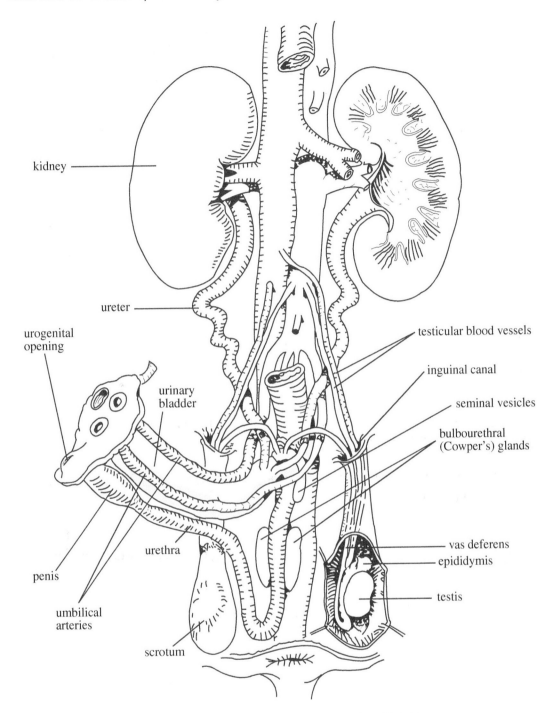

FIGURE 15-1. Urogenital System of the Male Pig

You can now see where the bladder and urethra are joined. Follow the urethra until it disappears beneath the **pubic symphysis** (midway between the two hip bones). The urethra will reappear in the region of the scrotum just ventral to the anus.

4. Slide the string out from beneath the tray **between the hind legs only.**

   Lift the posterior end of the pig. **While holding the posterior end elevated,** insert the scissors into the **anterior end** of the hip bones (the **pelvis**).

   Keeping the tip of your scissors pointing up, make a cut **about an inch** long through the pelvic bones.

   Bend the hind legs backwards until the pelvis cracks. The entire length of the urethra should now be visible (see **Figure 15-1**).

5. Carefully separate the urethra from the rectum by lifting and cutting below the urethra.

   This will expose the **three male accessory glands** (see **Figure15-2**). During ejaculation, these glands release the **seminal fluid** which mixes with and transports the sperm.

   A pair of large glands, the **bulbourethral (Cowper's) glands,** are visible on both sides of the urethra **near the anus.** These can be seen more clearly by lifting the urethra.

   Anterior to the bulbourethral glands, you will see a pair of small glands, the **seminal vesicles,** on the dorsal surface of the urethra.

   Between the two seminal vesicles, you will see an even smaller gland, the **prostate.** This gland may be difficult to see, since the pigs are small.

6. Return the urethra to its original position and look into the abdominal cavity. Toward the anterior end of the bladder, you can observe two small tubes (the **vas deferens**), which transport sperm from the testes to the urethra (see **Figure 15-1**).

   Trace the vas deferens posteriorly until it passes through an opening in the abdominal wall. The openings, called **inguinal canals,** are formed during the descent of the testes from the abdominal cavity to the **scrotum.** The scrotum is an external pouch of skin that contains the testes.

   If the inguinal canal is damaged or weakened, a loop of the intestine may penetrate and be pinched in the opening of the inguinal canal. This is called an inguinal hernia and must be surgically repaired.

   The **temperature of the testes is regulated** by the relative position of the scrotum to the body wall. In humans, normal sperm production requires a temperature about three degrees Centigrade lower than normal body temperature.

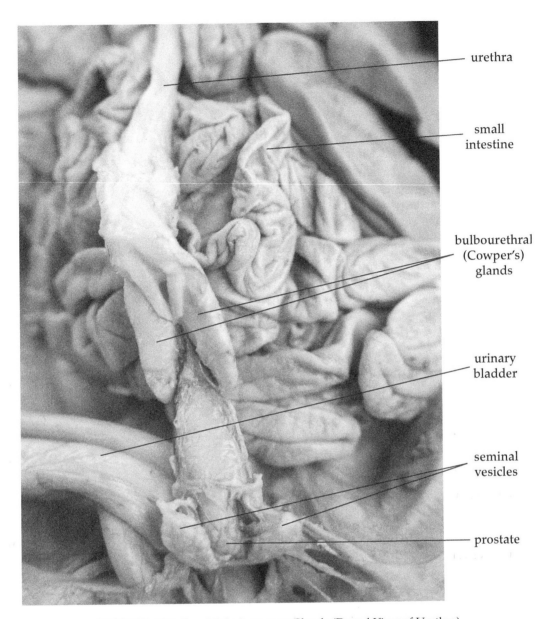

urethra

small
intestine

bulbourethral
(Cowper's)
glands

urinary
bladder

seminal
vesicles

prostate

FIGURE 15-2. Male Accessory Glands (Dorsal View of Urethra)

Optimal sperm production requires a temperature that is a few degrees lower than normal body temperature. In cold temperatures, the testes are drawn closer to the body cavity; exposure to warmth reverses the process.

Pull gently on one vas deferens and note the movement of the attached **testis** within the **scrotum.**

7. Carefully remove the membranes covering one testis. This will expose several structures.

    The **testis** (where sperm are produced by meiosis) appears as an oval. Note that this region is well supplied with blood vessels.

    Sperm production occurs in the **seminiferous tubules**, microscopic structures within the testes. The testes also produce male sex hormones, such as testosterone.

8. Along one edge of the testis is the **epididymis**, where sperm complete their maturation and are stored until ejaculation.

    Extending from the epididymis is a small tube — this is the **vas deferens** you were previously viewing from the abdominal cavity.

✓ Comprehension Check ✓

1. In the correct order, list each organ and/or structure a sperm passes through from the site of production to the outside of the body.

    Seminiferous tubulers, rete testis, vasa efferentia, Epididymis, vas deferens, Ejaculatory duct, Urethra, urethral meatus

2. Name **two** substances that travel through the **urethra**. What **two body systems** share this structure?

    Urian, Sperm / male urinary and reproductive system.

3. Sperm are stored in the ___*Epididymis*___ to complete their development.

4. In about 3% of male children, the testes fail to descend into the scrotum. Would this affect sperm production? **Explain** your answer.

    NO the system still works.

5. If a male has a vasectomy (an operation that closes the vas deferens), will he be able to ejaculate normally? If so, where is the ejaculated fluid produced and what is missing from the ejaculate? **Explain** your answer.

Yes You still make Semen, the Semen dosent contain Sperm

Check your answers with your instructor before you continue.

Dissection of a Female Fetal Pig

1. If the small intestine has not been removed from the abdominal cavity, **retract it anteriorly** to expose the **ovaries** (small oval structures on each side of the rectum, posterior to the kidneys).

   Inside the ovary are microscopic **follicles** that contain developing egg cells. The ovaries also produce the female sex hormones estrogen and progesterone.

   As you continue your dissection, refer to **Figure 15-3** to see the positions of the various organs.

2. There are two tubes positioned close to the ovaries. The smaller tube is the **oviduct (fallopian tube),** which **transports ovulated eggs** to the uterus and is where **fertilization** occurs.

   In the pig, the uterus is shaped like the letter **Y.** The larger, twisted tubes visible next to the ovaries are the **horns of the uterus.** In animals that have litters of young, uterine horns provide space for multiple embryos to develop.

   The uterine horns join together into the **body of the uterus** (see **Figure 15-3**), close to the anterior end of the bladder. In humans, the body of the uterus is larger, and usually accommodates a single embryo.

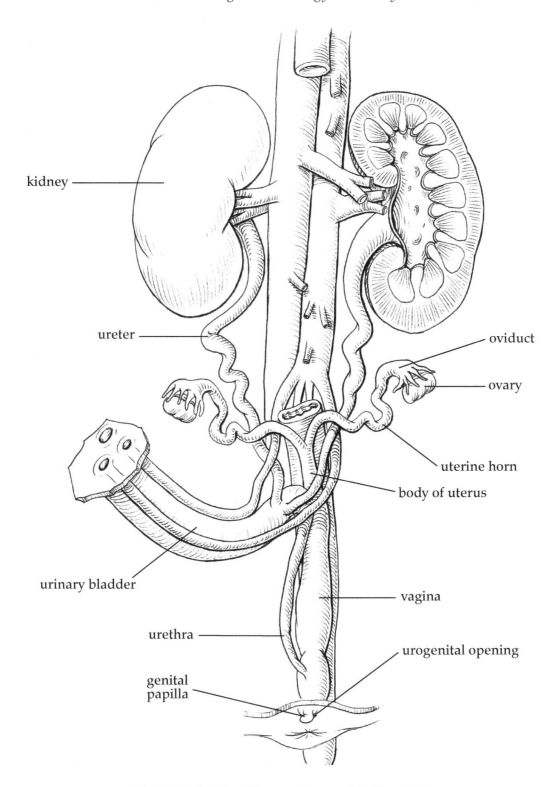

FIGURE 15-3.   Urogenital System of the Female Pig

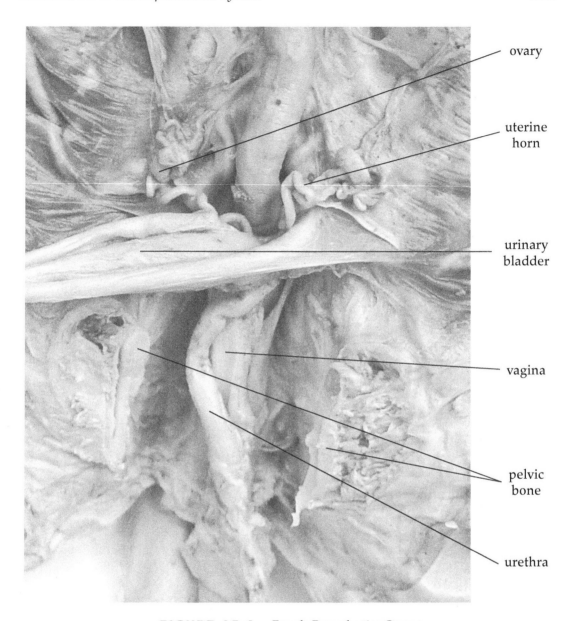

FIGURE 15-4. Female Reproductive Organs

3. Using scissors, separate the skin from the **urinary bladder** (which is located between the two umbilical arteries), by cutting posteriorally toward the anus. Don't remove the bladder.

   The **urinary bladder** is a **temporary storage organ for urine.** When the bladder fills, nerve endings in its muscular walls are stimulated, causing the muscles to contract for urination.

   You can now see where the bladder and urethra are joined. Follow the urethra until it disappears beneath the **public symphysis** (midway between the two hip bones).

4. Slide the string out from underneath the tray **between the hind legs only**.

Lift the posterior end of the pig. **While holding the posterior end elevated,** insert the scissors into the **anterior end** of the hip bones (the **pelvis**).

Keeping the tip of your scissors pointing up, make a cut **about an inch** long through the pelvic bones.

Bend the hind legs backwards until the pelvis cracks. The urethra will be visible, but not completely exposed.

5. Cut through the skin and muscle on both sides of the urethra, proceeding posteriorally to the urogenital opening.

You'll see two parallel tubes. The more **ventral** of the two is the **urethra.** The urethra **transports urine** out of the body through the **urogenital opening.**

Dorsal to the urethra you'll see the **vagina** (see **Figure 15-4**). The vagina is the muscular tube into which **sperm** are **deposited** during sexual intercourse, and also serves as the **birth canal.**

6. Insert the scissors between the vagina and the rectum and carefully cut in an anterior direction until the vagina separates from the rectum. Insert the scissors between the urethra and the vagina and carefully expose both tubes.

The **urethra and vagina join** together to form a common passage that leads to the **urogenital opening.**

7. On the external surface of the pig surrounding the urogenital opening, **you'll** see the **labia** (folds of skin). Gently spread the labia apart to see the tiny **clitoris** (erectile tissue).

✓ Comprehension Check ✓

1. In the correct order, list each organ and/or structure an egg will pass through from its site of production through fertilization, fetal development, and birth.

germinal, embryonic, and fetal stages

2. The female pig has a common opening for the urinary and reproductive systems. Do human females have the same anatomical pattern? If not, **explain the differences.**

    difference. The porcine uterus differs from the Human by being bicornuat.

3. If a woman has a tubal ligation (an operation that closes the oviducts), will she be able to ovulate normally? **Explain** your answer.

    Yes, it doesn't prevent ovulation or menstruation. Its known as female sterilization.

4. How does fertilization differ from implantation? Explain your answer, including the location where each takes place.

    Implantation happens outside of the fallopian tube

Check your answers with your instructor before you continue.

ACTIVITY 2   DEVELOPMENT OF GAMETES IN THE TESTES

In humans, sperm production begins at puberty and continues throughout a man's adult life. Sperm production **(spermatogenesis)** occurs by **meiosis** in the testes. Each ejaculation contains about 250 million sperm, so there is a continuous demand for large quantities of new sperm cells.

Sperm are formed within the seminiferous tubules of the testes. There are more than 100 meters of tightly coiled tubules in each testis. In the following activity, you will observe the development of sperm cells within the **seminiferous tubules** with the compound microscope.

1. Get a **compound microscope and a cross section slide of the testis.**

2. Observe the slide with the scanning lens **(4x)**. You'll see numerous circular structures. Each is a section through a **seminiferous tubule** (see **Figure 15-5**).

    Switch to **10x** objective and note that there are clusters of cells **between** the seminiferous tubules. These are referred to as **interstitial cells.** They produce **male sex hormones (androgens),** the best known of which is **testosterone.**

3. Observe the slide on high power **(40X)** to get a closer view of the developing cells within **one tubule** (as shown in **Figure 15-6**).

    You'll see that the tubule is packed with cells in various stages of development. Spermatogenesis begins at the outermost layer of the tubule.

    As repeated cell divisions occur, the developing cells move closer and closer to the center of the tubule. Therefore, your observations will also start near the outer wall of the tubule and move progressively inward toward the **lumen** (center) of the tubule (see **Figure 15-5**).

4. Begin your observations close to the outer wall of the tubule. You're looking for tiny cells that lie directly adjacent to the tubule wall. They're usually darker in color than other cells of the tubule. These are the **spermatagonia,** the stem cells that give rise to sperm.

    Refer to **Figure 15-6** for a correlation between the various stages of meiosis with the position of the developing sperm within the seminiferous tubules.

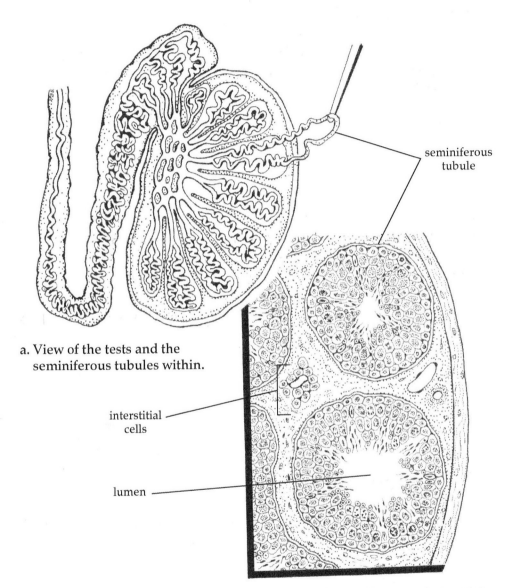

a. View of the tests and the
   seminiferous tubules within.

seminiferous
tubule

interstitial
cells

lumen

b. Cross section of seminiferous tubules (4X).

FIGURE 15-5.   Spermatogenesis within the Seminiferous Tubules

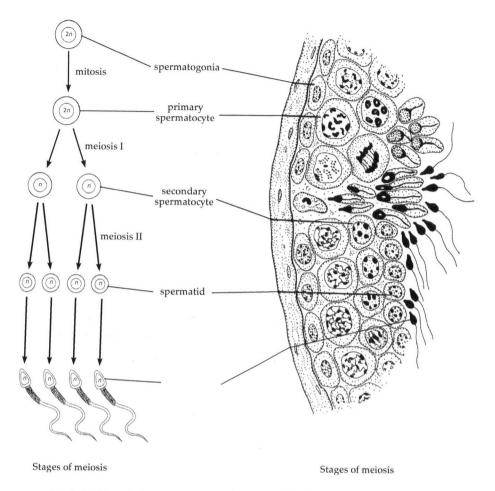

FIGURE 15-6.   Gamete Production within the Seminiferous Tubules.

5. The **spermatogonia divide by mitosis,** producing two daughter cells with each division. Some of these daughter cells remain as spermatogonia and others develop into **primary spermatocytes.** The primary spermatocytes begin **meiosis I,** which produces **two secondary spermatocytes.**

Look for the **primary spermatocytes** in a layer just inside that of the spermatagonia (closer to the lumen). The primary spermatocytes are larger than the spermatagonia.

The next layer contains **secondary spermatocytes,** which are undergoing meiosis II. Meiosis II produces four daughter cells called **spermatids.** Each is **haploid,** containing only **23 chromosomes** (half the number of a normal body cell).

The spermatids differentiate into **sperm.** The tails of many can be seen projecting into the lumen of the tubule. Immature sperm continue their maturation in the epididymis, which lies just outside the testis.

✓ Comprehension Check ✓

1. A body cell in a house cat has 38 chromosomes. If we were looking at the testes of a male house cat, how many chromosomes would be in a **spermatid?**

    How many chromosomes would be in the **spermatogonia?**

    4 8 chromosomes.

    19 spermatid

2. The process of spermatogenesis and maturation takes approximately nine weeks to complete. If two ejaculations occur within a short time period, will the second ejaculation contain sperm? **Explain** your answer.

    Yes, you are always repuducing sperm.

3. Note that there are many more spermatids than secondary spermatocytes in each tubule. **Explain** why.

    The Secondary Spermatocyte gives birth to Spermatids after meiosis II

Check your answers with your instructor before you continue.

ACTIVITY 3   DEVELOPMENT OF EGGS IN THE OVARIES

In humans, egg production (**oogenesis**) occurs by **meiosis** in the follicles of the ovaries. At puberty, a woman has approximately 400,000 follicles. Most of the eggs inside these follicles, however, will never be ovulated, since usually only one egg is released per menstrual cycle.

In the following activity, you will observe the development of egg cells with the compound microscope.

1. Work with a group. Get **two compound microscopes, one dissecting microscope, and three cross section slides of the ovary: immature follicles, mature follicles, and corpus luteum.**

2. Position the two compound microscopes side by side on the laboratory table. Place the slides of **immature and mature follicles** on the stages of the microscopes.

   Observe the slide of the immature follicles with the scanning lens (**4X**). Close to the **outer edge** of the ovary, you'll see clusters of small follicles (see **Figure 15-7a**). Inside each follicle is a **primary oocyte**, a developing egg. The primary oocyte surrounded by follicle cells is called a **primordial follicle**.

   Switch to the **10X** objective for a closer view of the primordial follicles.

   After puberty, under the influence of hormones, various groups of primordial follicles are activated and continue the stages of development outlined below.

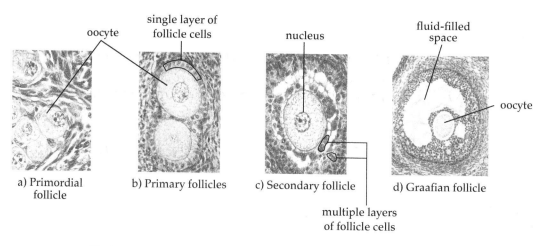

FIGURE 15-7. Stages of Follicle Development within the Ovary

3. Near the clusters of primordial follicles, you will observe larger follicles, each with only a **single layer** of follicle cells surrounding the oocyte, which has also increased in size (see **Figure 15-7b**). These are called **primary follicles.**

4. Locate a follicle with **several distinct layers of follicular cells** surrounding the oocyte. Switch to the **(10X)** objective for a closer view (see **Figure 15-7c**).

    These are the **secondary follicles.** As secondary follicles mature, spaces appear within the follicular layers. As you scan the slide, you'll see secondary follicles in various stages of development.

    Cells of developing follicles produce the hormone **estrogen.** As the layers of follicle cells thicken, estrogen secretion increases.

    Move to the **second slide (mature follicles).**

5. Locate a **Graafian follicle** on the **mature follicle slide** (see **Figure 15-7d**). The Graffian follicles are quite **large, with a fluid-filled space** between the egg and the surrounding follicular cells (since this fluid is usually lost in slide preparation, the space will appear empty). The egg is often located on a small mound of follicle cells.

    Switch back and forth between the two microscopes, comparing the structure and appearance of the primary and secondary follicles with the Graafian follicles.

    Near the end of the development of the Graafian follicle, the primary oocyte inside the follicle completes meiosis I and begins meiosis II. Each daughter cell of meiosis is **haploid,** containing only **23 chromosomes** (half the number of a normal body cell).

    The Graafian follicle releases the egg and **ovulation** occurs. In humans, meiosis II only occurs after an egg has been fertilized.

Note:

The slides that are used to demonstrate oocyte development are often obtained from animals that produce multiple young. Therefore, more follicles will be present than would be observed in human ovaries.

---

6. Place the slide of the **corpus luteum** on the stage of the **dissecting microscope.**

    The corpus luteum is an enormous circular or oval mass of cells. It may appear to form a bulge on the side of the ovary.

    The corpus luteum forms from the empty follicle remaining after ovulation. The cells of the corpus luteum produce the hormone **progesterone,** which prepares the uterus for pregnancy. The corpus luteum also continues to produce small amounts of estrogen, but progesterone is the primary hormone produced by this structure. If the egg is fertilized the corpus luteum continues to function for several months during pregnancy. If fertilization does not occur, however, the corpus luteum gradually degenerates, and progesterone production declines.

✓ Comprehension Check

1. A human body cell has 46 chromosomes. If we were looking at an **ovulated egg that has been fertilized,** how many chromosomes would there be from the original egg cell?

   How many total chromosomes would be present (from both the egg and sperm)?

   23

2. Which would produce more estrogen: a primordial follicle or a secondary follicle? **Explain** your answer.

   Secondary because the primary follicle developes into the secondary one.

3. If ovulation occurs at approximately midpoint in the reproductive cycle, would you expect to find secondary follicles before or after ovulation?

   Would you expect to find a corpus luteum before or after ovulation?

   Explain your answers.

   after ovulation because the primary follicles are developed into secondary ones

4. **Challenge question!** On a slide of the ovary, it is common to see large follicles with no eggs visible. Suggest an explanation for this observation.

   was not fertilized.

Check your answers with your instructor before you continue.

Self Test ✓

1. Fill in the blanks with the name of the appropriate reproductive structure or its function.

STRUCTURE	FUNCTION
Ovaries,	production of sperm and sex hormones
follicle	Plays a major role in the dual function of the ovary.
seminiferous tubules	maintain sperm, store
testies	production of eggs and sex hormones
Scrutum.	pouches of skin outside the abdomen that enclose the testes; provide a cooler environment for sperm development
epididymis	a long tube that is located near each testicle
uterus	site of embryo implantation and fetal development
Semen	production of lubricating mucus and secretion of seminal fluid; provides a fluid transportation medium for sperm
vas deferens	the duct which conveys sperm from the testicle to the urethra
vagina	muscular tube that receives sperm; also serves as the birth canal
oviduct (fallopian tube)	Produce eggs and hormones
urethra	muscular tube that carries urine from the urinary bladder to the outside; in males, also conveys semen during ejaculation
penis	urination and sexual intercourse

2.   Correctly label the following structures on **Figure 15-8**: urethra, vas deferens, ~~testis, inguinal canal, umbilical arteries, pelvic bones,~~ and ~~urinary bladder.~~

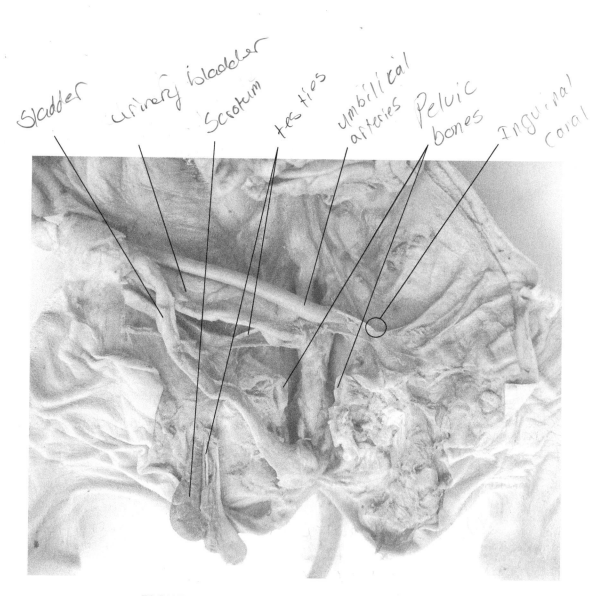

FIGURE 15-8.   Reproductive System of the Male Pig

3. Correctly label the following structures on **Figure 15-9**: ~~urethra~~, ~~vagina~~, ~~uterine horns~~, ~~umbilical arteries~~, ~~ovary~~, ~~pelvic bones~~, ~~urinary bladder~~, and ~~genital papilla~~.

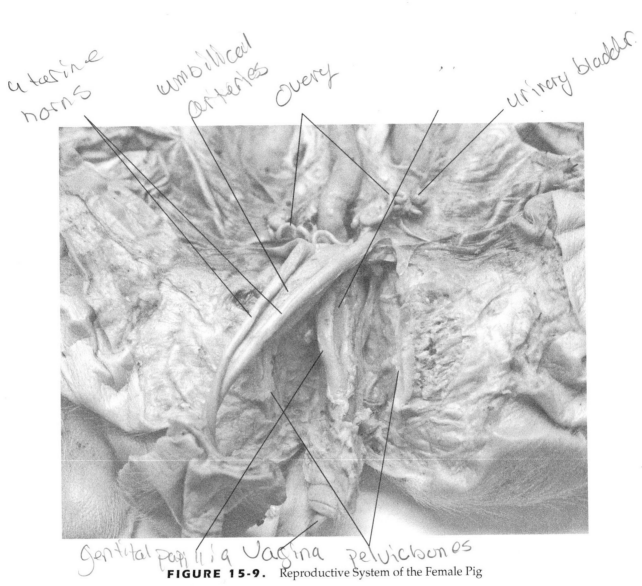

uterine horns

umbilical arteries

Ovary

urinary bladder

Gentital papillia   Vagina   Pelvic bones

Urethra

FIGURE 15-9. Reproductive System of the Female Pig

4. Explain what happens to a ruptured follicle after ovulation.

developes corpus luteum.

5. If a woman has to have her uterus removed, her ovaries might be left in position because:

Hormonal balance
Stucture.

6. A fertility clinic in Germany has been advising its male clients to avoid wearing excessively tight jeans. Is there any physiological basis for this recommendation? **Explain** your answer.

Yes over heat testies,

7. How does the function of the urethra in males differ from its function in females? **Explain** your answer.

males use it for uration and ejaclation. where females use it for unary.

8. Explain why you would not expect to find a corpus luteum prior to puberty.

Because eggs havent drop,

16

Mitosis and Asexual Reproduction

Objectives

After completing this exercise, you should be able to:

- name and describe the stages of mitosis
- identify the stages of plant and animal mitosis as viewed through the compound microscope
- discuss the relationship between the number of cells in each of the stages of mitosis and the length of the various stages
- explain the relationship between mitosis and the processes of regeneration and asexual reproduction
- apply your knowledge of asexual reproduction to cloning and other medical technologies

CONTENT FOCUS

Most of us are aware that the outer layer of our skin is subject to constant wear. To keep this protective layer intact, skin cells must be replaced throughout our lives.

Cells of the epidermis, for example, are replaced every 25–45 days. A similar situation occurs elsewhere in the body.

Growth is another activity for which additional body cells are required. A **zygote** (fertilized egg) begins life as a single cell that multiplies into many cells as the embryo develops. During childhood, cell multiplication provides the many cells needed for growing into an adult.

The type of **cell division** that makes growth and repair possible is called **mitosis.** All body cells produced by mitosis must contain **the same genetic information** as all other body cells — the information that makes you a unique individual.

ACTIVITY 1   HOW MITOSIS WORKS

A cell is genetically "programmed" to carry out its function in the body. These cellular instructions are included within structures referred to as **chromosomes.** A full set of these genetic instructions is necessary if a cell is to function normally. Therefore, when new body cells are produced by **mitosis,** each has a complete set of chromosomes.

The normal number of chromosomes in a cell is referred to as the **diploid number (abbreviated 2n).** Each type of organism has a characteristic number of chromosomes. The diploid number of chromosomes in human body cells, for example, is 46. The diploid number in one species of pine tree is 24, in crayfish it's 200, and in fruit flies it's only 4.

Assume that the diploid number of chromosomes in the cell below equals 16.

If this cell divides, how many chromosomes must be present in each new cell?

Enter this number in each of the blank circles below.

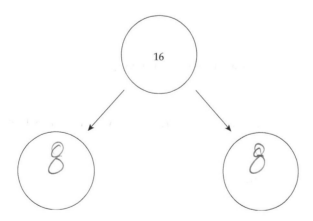

To make it easier to understand, the process of cell division is divided into five stages (called the **cell cycle**). At the completion of the cell cycle, two new cells are formed. Each cell will have a complete set of genetic instructions.

To complete cell division successfully, **dividing cells must solve several problems:**

■ **Each new cell needs a full set of chromosomes.**

Accomplished by: Duplicating chromosomes (an exact copy is produced).

To avoid confusion, these duplicate chromosomes are referred to as **sister chromatids.**

A cell preparing to divide will contain two complete sets of sister chromatids (one set for delivery to each of the two new cells).

■ **Sister chromatids must be attached together.**

Accomplished by: A structure called a **centromere** fastens the duplicates together.

Attachment makes it easier to keep track of sister chromatids for sorting into two groups (one set for each new cell).

■ **Pieces of chromosomes shouldn't get broken off or lost.**

Accomplished by: The long, threadlike chromosomes coil and fold into compact structures. Condensed chromosomes are thick enough to be visible with the compound microscope.

The pairs of sister chromatids must be pulled apart and delivered to each new cell.

Accomplished by: A structure called the **mitotic spindle** hooks on to each pair of sister chromatids. **Spindle fibers** pull the chromatids apart and deliver one chromatid from each pair to each new cell.

■ **When the sorting and delivery process is complete, the two cells must be separated into two daughter cells, each with its own set of chromosomes.**

Accomplished by: A process called **cytokinesis** separates the cytoplasm into two halves (**"cyto"** means *cell* and **"kinesis"** means *cutting*).

During cytokinesis in animal cells, the cell membrane pinches into a groove called the **cleavage furrow.**

ACTIVITY 2   RECOGNIZING THE STAGES OF MITOSIS

Use the **Decision Tree in Figure 16-1** to **identify the stages of the cell cycle** in **Figure 16-2.**

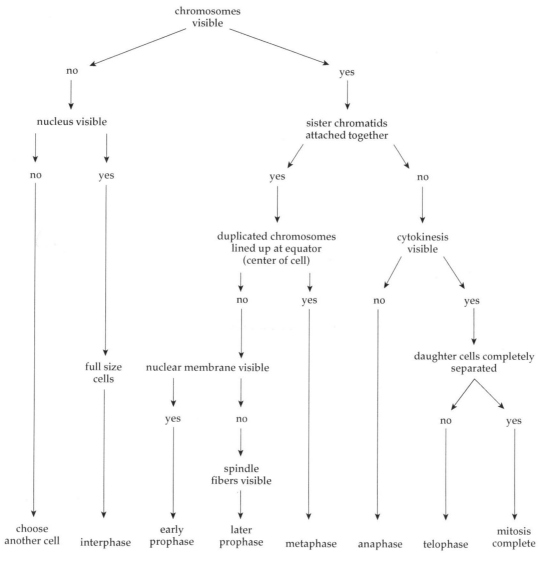

FIGURE 16-1. Mitosis Decision Tree

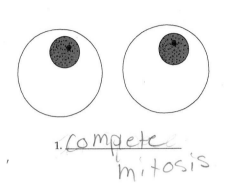

1. Complete mitosis

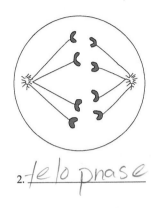

2. telo phase

3. Later Prophase

4. metaphase

5. Interphase

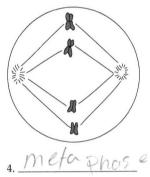

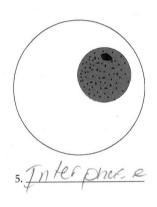

6. prophase

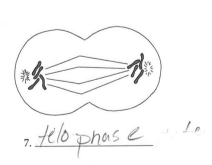

7. telophase ...le.

FIGURE 16-2. Stages of the Cell Cycle to be Identified

✓ Comprehension Check

1. Arrange the following stages of the cell cycle in the correct order (1 through 5).

    __3__ metaphase   __5__ telophase

    __1__ interphase   __2__ prophase

    __4__ anaphase

    Use the following words to fill in the blanks in the questions below. Each answer can be used **only once**.

    centromere   sister chromatids   cleavage furrow
    equator   cytokinesis   chromosomes
    mitotic spindle

2. During metaphase, chromosomes line up along the _equator_ of the cell.

3. The two _sister chromatide_ are held together with a _Centromere_.

4. _Cytokines_ refers to the division of the cytoplasm into two daughter cells.

5. Genetic information in cells is contained in structures called _Chromosomes_.

6. During anaphase, part of the _Mitoticspindle_ can be seen attached to the centromere of each sister chromatid.

7. The constricted appearance of the cell membrane prior to the formation of two daughter cells is called the _cleavag frournew_.

Check your answers with your instructor before you continue.

ACTIVITY 3   MITOSIS — THE REAL THING!

The stages of the cell cycle in living cells are often not as clear as in a set of diagrams. Sharpen your observation skills and practice recognizing the stages of mitosis at the same time.

1. For this activity **work on your own.** Get the following supplies: **one slide of mitosis in whitefish embryos and a compound microscope.**

2. View the slide with the **high-power lens (40×).**

3. Your instructor will assign you **one stage of the cell cycle** to locate on your slide.

    When you locate a **clear** example of that stage, **indicate the cell with the pointer** in the ocular lens.

 Check your identification with your instructor before you continue.

4. Draw the stage **as it appears under the microscope,** in the appropriate location on **Figure 16-3.**

    In your drawing, label the following (if present): **chromosomes, mitotic spindle, equator, cleavage furrow, cell membrane, and nuclear membrane.**

5. After completing your drawing, **observe** the other **stages of the cell cycle** as identified by other students.

    Create your own drawings of these stages and add them to the appropriate locations on **Figure 16-3.**

    In your drawings, label the following (if present): **chromosomes, mitotic spindle, equator, cleavage furrow, cell membrane, and nuclear membrane.**

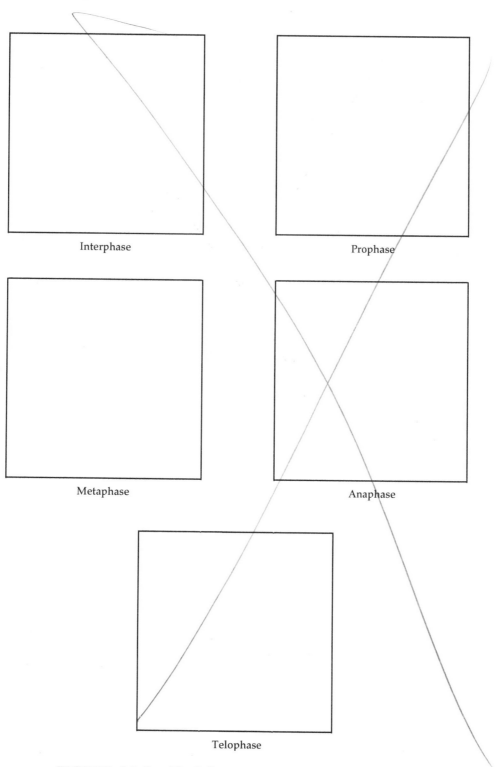

FIGURE 16-3. The Cell Cycle as Seen through the Microscope

ACTIVITY 4

ESTIMATE THE DURATION
OF THE CELL CYCLE

When you look at a prepared slide, you're looking at a "moment frozen in time." The preserving chemicals stopped the cells in the middle of their daily activities. From this slide, you can see how many cells were undergoing cell division at that moment. You can even see how many cells were in each of the stages of mitosis. You can use this information to estimate the length of each stage of the cycle compared to the other stages.

1. For this activity, **work individually.** Get the following supplies: **one slide of onion root mitosis and a compound microscope.**

   View the slide with the **high-power lens.** Position your view of the onion root tip to a section where many cells are undergoing mitosis (see **Figure 16-4** for examples).

   Your instructor will give you additional instructions on how to position the slide for optimal viewing.

2. Using the stages of onion root tip mitosis in **Figure 16-4** as a guide, **count** the number of cells **in your field of view** that are in each stage of the cell cycle.

   Record your results in **Table 16-1.**

TABLE 16-1
RESULTS OF CELL CYCLE OBSERVATIONS

NUMBER OF CELLS IN STAGE	INTERPHASE	PROPHASE	METAPHASE	ANAPHASE	TELOPHASE
View 1	1	1	1	1	2
View 2					
View 3					
Total for views 1–3					
Percentage of cell cycle	20 %	40 %	60 %	80 %	100 %

Interphase   Prophase   Metaphase   Anaphase   Telophase

FIGURE 16-4.   Stages of Mitosis in the Onion Root Tip

3. Repeat the **instructions in Step 2 twice more** (so you'll examine a total of **three different onion fields of view**) and **record the results** for each in **Table 16-1.**

4. **Total the number of cells counted** for each view and record the totals in **Table 16-1.**

5. **Calculate** the percentage of cells observed in **each stage of the cell cycle.**

    STEP 1:

    Add the total number of cells counted in all five stages.

    Total cells counted = interphase + prophase + metaphase + anaphase

    + telophase

    Total cells counted = ___2___

    STEP 2:

    Divide the number of cells in interphase by the total from Step 1 and multiply by 100.

    $$\frac{\text{Number in interphase}}{\text{Total cells counted}} \times 100 = \underline{\quad 50 \quad} \%$$

    STEP 3:

    Repeat **Step 2** for the remaining four stages of the cell cycle. **Record all percentages** in **Table 16-1.**

6. Using the information in **Table 16-1,** answer the following questions about the **length** of each stage of the cell cycle.

Which do you think is the longest stage? _____Prophase_____

Which do you think is the shortest stage? _____Anaphase_____

Do you think any of the stages are about the same length? If so, list them below.

Interphase

metaphase

telophase

7. See **Table 16-2 (after the Self Test)** for the results of some experiments on the length of the onion cell cycle. Do the experimental results support your conclusions about the length of the various stages of mitosis? _____Yes_____ **Explain** your answer.

Because we did only one.

ACTIVITY 5   REGENERATION

Some animals have the ability to replace (regrow) lost or damaged body parts. This process is known as **regeneration** and is a form of asexual reproduction. You may be familiar with some animals that have this ability. Starfish (also known as **sea stars**) can regenerate body parts as long as a small part of the central disk is present. Sea stars can grow a new arm and a severed arm can grow a whole new body! If cut in two, each half of the sea star can develop into a whole individual.

During the microscope exercise, you looked at planaria, freshwater flatworms. Planaria also have the ability to regenerate lost body parts. Any piece of the body except the tail can regenerate a complete new adult. In this activity, you'll cut **planaria** to demonstrate regeneration and regrowth.

1. Work in groups. Get the following supplies: **one clean razor blade in a plastic holder, a glass petri dish, a pipette, and crushed ice.**

2. You'll also need a bottle of **pond water, labeling tape, and a dissecting microscope.**

    Set up your dissecting microscope, but keep the light **off.** On the stage of the microscope, place the **top half** of the petri dish.

    Fill the dish **to the rim** with crushed ice.

3. Place **one large planaria** into the **bottom half** of your petri dish with a small amount of **pond water.**

    Set the dish with the planaria **onto the crushed ice.** This will anesthetize the worm for the "operation."

    Let the worm relax on the ice for about **five minutes.**

4. While you're waiting, decide how you'll cut the planaria. You can choose one of the two cutting patterns in **Figure 16-5,** or develop your own pattern.

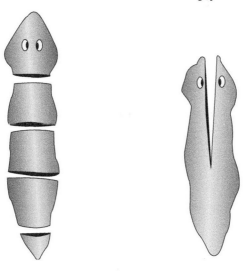

FIGURE 16-5. Planaria Cutting Patterns

5. With the razor blade, make **perpendicular cuts** through the worm. Don't tilt the razor blade when cutting. The cuts must be sharp and clean.

6. After the operation, add some clean **pond water** to the petri dish containing the worm.

   Discard the ice. Wash and dry the top half of the petri dish. Using labeling tape, **mark the lid** with **your laboratory section and your name.**

   Return the petri dish containing the worm to your instructor to be stored in a cool, quiet location.

7. After **two weeks,** you can examine your worm under the dissecting microscope to see the results of your experiment.

8. **Challenge Question!** If you cut a planaria into five pieces and each piece developed into a separate individual, would you be able to tell them apart? **Explain** your answer.

ACTIVITY 6   ASEXUAL REPRODUCTION: GROWING A PLANT FROM A CUTTING

As you've seen, mitosis provides genetically identical cells for growth, replacement, and repair of body tissues. Mitosis also provides a mechanism for **reproducing whole organisms asexually.** In asexual reproduction, offspring are produced which are **genetically identical** to the single parent and also to one another.

Plants don't have to grow from seeds. You can remove cuttings from your plants. The cuttings will develop roots and grow into new plants. This form of asexual reproduction is an example of **plant propagation.**

1. Get **a paper cup and a scalpel.**

   You'll also find **plants for propagation, rooting hormone, and potting soil.**

2. Fill the cup with **potting soil** up to **one inch** from the top of the container.

   Moisten the soil with water until it's damp, but not soggy.

3. **Cut a small piece of stem** with several leaves attached.

   Dip the **cut end** of the stem into the **rooting hormone** and **plant** it in the cup.

4. Keep your young plant **moist** and exposed to **bright, indirect light.** Roots should form within several weeks.

✓ Comprehension Check

1. A botanist is trying to save a rare plant from extinction. Suggest a method she can use to increase the population.

2. The **type of cell division** that produced root growth in my cutting is

   _____ .

3. The possibility of human cloning, a type of asexual reproduction, is often in the news these days. If a baby could be cloned using the DNA from one of her mother's skin cells, what percent of the baby's traits would be similar to those of her mother? **Explain** your answer.

4. How would a colony of mice **produced by cloning be helpful** for testing the effectiveness of a new diet pill?

Check your answers with your instructor before you continue.

Self Test

Fill in the blanks with the most appropriate answer. Answers can be used **only once.**

a. diploid
b. sister chromatid
c. centromere
d. spindle fibers
e. nuclear membrane
f. cell membrane

g. cytokinesis
h. cleavage furrow
i. equator
j. zygote
k. daughter cell
l. chromosome

1. _____ Breaks down at the end of prophase, releasing the chromosomes.

2. _____ Chromosomes are arranged along this imaginary line during metaphase.

3. _____ Structure that holds two duplicate chromosomes together.

4. _____ Two of these are found at the **end** of mitosis.

5. _____ Normal number of chromosomes in a cell.

6. _____ Responsible for chromosome movement during mitosis.

7. _____ A constriction of the cell membrane that occurs in telophase.

8. _____ Duplicated chromosome has two of these.

9. _____ Process that separates the cytoplasm into two halves.

10. _____ Genetic material found in the cell nucleus.

Fill in the blanks with **true** or **false.**

11. _____ If a cell undergoing **mitosis** has a chromosome number of six, the daughter cells will have a chromosome number of three.

12. _____ In asexual reproduction, the offspring are identical to the parent.

13. _____ If cells are multiplying, as they do when grown in tissue cultures, mitosis is taking place.

14. _____ Growth during childhood is a good example of cell division.

15. Complete mitosis for the following parent cell. Draw each stage of mitosis **in order,** and show the number of chromosomes in each stage (including the two daughter cells). The beginning cell has a diploid number of four chromosomes.

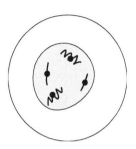

16. You're vacationing in Maine. One morning, while walking near the shore, you hear a group of fisherman complaining that there are too many sea stars in the ocean. They plan to solve this problem by netting the sea stars, cutting them in half to kill them, and throwing them back into the ocean. Is their plan likely to be effective? **Explain** your answer.

17. **Challenge Question!** If you wanted to asexually propagate a human, you could use

    a. a skin cell
    b. a sperm cell
    c. a red blood cell
    d. any of the above would be suitable
    e. none of the above would work out

    Explain your answer.

TABLE 16-2 RESULTS OF EXPERIMENTS ON THE LENGTH OF STAGES IN THE ONION CELL CYCLE					
	INTERPHASE	PROPHASE	METAPHASE	ANAPHASE	TELOPHASE
Length of stage	17 h	88 min	4 min	3 min	6 min
Percentage of cell cycle	91%	7.9%	0.4%	0.3%	0.5%

Connecting Meiosis and Genetics

Objectives

After completing this exercise, you should be able to:

- name and describe the stages of meiosis
- correctly use and understand the terminology associated with cell division and genetics
- demonstrate an understanding of the changes in chromosome number that occur during meiosis and fertilization
- compare meiosis I with meiosis II in terms of the position of the chromosomes in each stage, changes in chromosome number, and number of daughter cells produced
- explain the process and importance of crossing-over between homologous chromosomes
- draw and complete Punnett squares and use them to determine genetic probabilities in monohybrid crosses

CONTENT FOCUS

Your body cells contain **46 chromosomes** within the nucleus. As you probably know, **half** of this genetic information (23 chromosomes) is inherited from your mother and **half** (the other 23 chromosomes) from your father.

The genetic information carried by the **gametes** (the sperm and egg), when incorporated into a fertilized egg, will determine all the physical and physiological traits of the offspring.

There is a special type of cell division that changes the chromosome number from the normal **diploid** number (46) to the **haploid** number (23) found in sperm and eggs. This special type of cell division is called **meiosis.**

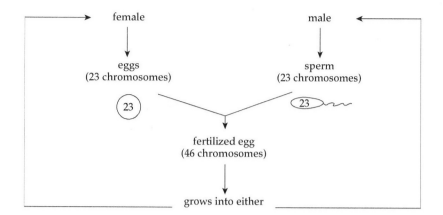

The stages of meiosis, in most respects, are similar to those of mitosis. The important differences include the following:

■ the chromosome number is reduced from diploid to haploid to form gametes

■ an exchange of genetic material takes place in a process known as **crossing-over;** gametes produced by meiosis are **not** genetically alike

■ meiosis involves two cell divisions (meiosis I and meiosis II); one occurs immediately after the other

The genes on a chromosome may exist in **more than one form,** called **alleles.** Individuals inherit **two alleles for each trait,** one received from the mother in the egg and the other from the father in the sperm.

If the two inherited alleles represent **different** forms of the gene (different **alleles**), an individual is called **heterozygous for that trait (Bb).** The allele that is **expressed in heterozygous individuals** is referred to as **dominant** and is represented by a capital letter **(B).**

The allele **not expressed** in heterozygous individuals is called **recessive.** Recessive alleles are represented with lowercase letters **(b).**

If individuals inherit two identical alleles for a trait **(BB or bb),** they are said to be **homozygous for that trait.**

The **combination of alleles** for a trait represents an individual's **genotype** (such as BB or Bb).

The **physical description** of a specific trait is called the **phenotype** (such as brown eyes, blood type, or freckles).

In this exercise, you'll work through the process of meiosis to **form** male and female **gametes, decode** the genetic information carried by these gametes, and use that information to **build** your baby's face.

GETTING STARTED:
ACTIVITY 1   BUILD A PAIR OF CHROMOSOMES

1. Work in groups. Get the following supplies: a **bag labeled "diploid human genome, male"** or **"diploid human genome, female."**

    In the supply area, you'll find containers of **beads** of several colors and shapes (representing **alleles**) and magnetic **connectors** (representing **centromeres**).

    Although humans have 23 pairs of chromosomes, we'll simplify the process by using only **one chromosome pair.**

2. Fill your bag with **four beads from each jar.** Note that there are two jars of beads of each color. In one jar, the beads are striped. In the other jar they're solid.

    Solid-colored beads represent **dominant alleles** and **striped beads** represent **recessive alleles for the same trait.**

    Get **two red centromeres and two yellow centromeres.**

3. Remove Table 17-1 from one group member's book and lay it flat on the laboratory table. As you draw each bead (representing the alleles of your maternal and paternal chromosomes), place it into the appropriate box in Table 17-1.

    Bead selection is random! Don't look in the bag while you're drawing the beads.

    a. The first bead you draw will become part of the chromosome you inherited from your **mother (maternal** chromosome).

    b. The next bead of the **same color** you draw will become part of the chromosome you inherited from your **father (paternal** chromosome).

    c. Additional beads of the **same color** you draw will be **THROWN BACK INTO THE BAG.**

    d. Draw beads until you've drawn **one of each color or shape for the maternal chromosome and one of each color or shape for the paternal chromosome.**

4. When you've drawn all the beads you need for both chromosomes, you're ready to hook your alleles together.

    Take a **yellow centromere** and hook it into position **between the green and red beads** of your **maternal chromosome.**

    Keeping the beads in the **correct order,** attach the remaining beads to the chromosome.

TABLE 17-1

TRAITS ON YOUR CHROMOSOMES

BEAD COLOR OR SHAPE	MATERNAL ALLELES	PATERNAL ALLELES	TRAIT	ALLELES (GENOTYPE)
Purple			Face shape	FF or Ff — round ff — triangular
Orange			Hair texture	HH — curly Hh — wavy hh — straight
Yellow			Eye size	EE — large Ee — medium ee — small
Blue			Eye distance	DD — close together Dd — medium spacing dd — far apart
Green			Eyebrow shape	BB or Bb — thick bb — thin
Red			Eyelash length	LL or Ll — long ll — short
White			Nose size	NN — big Nn — medium nn — small
Pink			Lips	GG or Gg — thick gg — thin
Black			Ear lobes	RR or Rr — free rr — attached
White oval			Cleft in chin	TT or Tt — present tt — absent
White twisted			Freckles	QQ or Qq — freckles qq — no freckles

5. **Repeat step 4** to connect the alleles of the **paternal chromosome** using the **red centromere.**

   Together these **two chromosomes (maternal and paternal)** are referred to as **homologous chromosomes.** Homologous chromosomes carry alleles for the **same traits** (face shape, eye size, etc.), although the genetic information is **not identical.**

6. Fill in **Table 17–2** with the genotypes and phenotypes of the traits on the chromosomes you've just constructed.

   Solid beads = dominant alleles. Striped beads = recessive alleles.

TABLE 17-2 WHAT ARE YOUR TRAITS?		
TRAIT	YOUR GENOTYPE	YOUR PHENOTYPE
Face shape		
Hair texture		
Eye size		
Eye distance		
Eyebrow shape		
Eyelash length		
Nose size		
Lips		
Ear lobes		
Cleft in chin		
Freckles		

Note:

When selecting your genotypes and phenotypes from Table 17-1, you probably noticed something unusual. Some of the traits have two possible phenotypes (round face or triangular face) and others have three options (curly hair, wavy hair, or straight hair).

The traits with three possible phenotypes represent a different mode of inheritance called incomplete dominance. Incomplete dominance is a special type of inheritance that occurs when an allele exerts only partial dominance over another allele. This results in a third, intermediate phenotype in heterozygous individuals.

	HOMOLOGOUS CHROMOSOMES
ACTIVITY 2	SEPARATE IN MEIOSIS TO FORM GAMETES

1. Draw a series of circles that are the same as those shown in **Figure 17-2.**

    Make the circles big enough for your chromosomes to fit comfortably.

2. Place your pair of chromosomes in the **first** circle. Your cell is now in **interphase.**

    What is the diploid number of chromosomes in ths cell? _____

    How many chromosomes came from your mother? _____

    How many from your father? _____

    How many **traits** are represented on your chromosomes? _____

    How many **alleles** are represented on your chromosomes? _____

3. An important event that occurs during **interphase** involves the replication of chromosomes. Use the spare beads in your genome bag to **replicate** your two chromosomes.

    Two identical (replicated) DNA strands are called **sister chromatids.**

    Attach the new **sister chromatids** for your **maternal chromosome** together at the centromeres. Do the same for your replicated **paternal chromosome.**

4. Move your duplicated chromosomes to the next circle, marked **prophase** through **telophase of meiosis I.**

    You'll complete **prophase through telophase of meiosis I** in the **same circle.**

5. As shown in **Figure 17-1,** during **prophase I,** the **two homologous chromosomes** find each other and pair up in a process called **synapsis.**

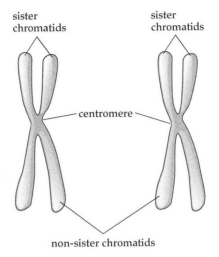

sister
chromatids

sister
chromatids

centromere

non-sister chromatids

FIGURE 17-1. Pattern for Meiosis Simulation

6. During synapsis, the **non-sister chromatids** exchange genetic information. This process is called **crossing-over.**

    Simulate crossing-over between **two non-sister chromatids** in your homologous pair (exchange of alleles between **one maternal and one paternal** chromosome).

    Exchange alleles for the **last five traits** (nose size through freckles).

Note:
Exchange alleles ONLY for two NON-SISTER chromatids!

7. Simulate **metaphase I** by placing the chromosomes **in the correct position** relative to the **equator.**

    To simulate **anaphase I,** separate the two homologous chromosomes by moving them to the opposite poles of the cell.

    To simulate the division of the cytoplasm in **telophase I,** draw a dotted line that represents the separated **daughter cells.**

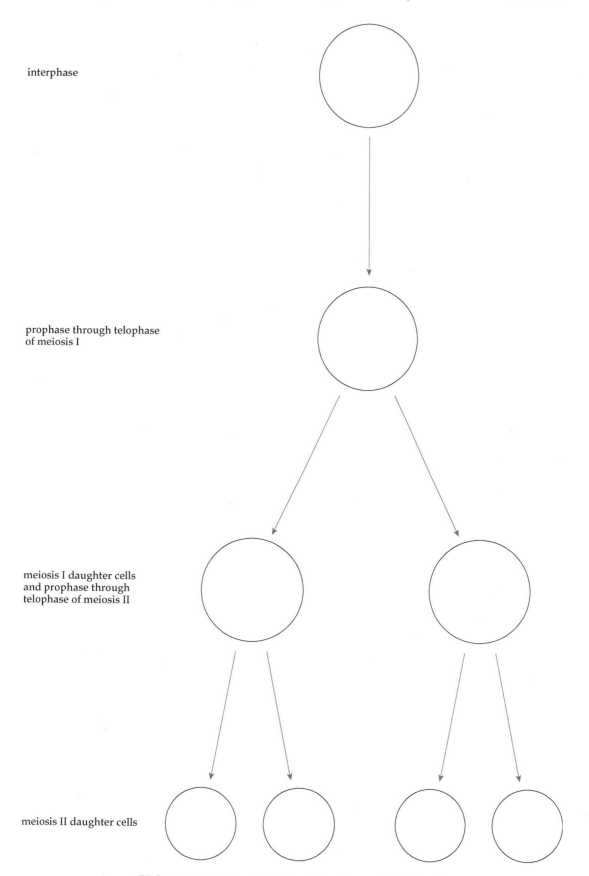

interphase

prophase through telophase
of meiosis I

meiosis I daughter cells
and prophase through
telophase of meiosis II

meiosis II daughter cells

FIGURE 17-2. Duplicated Homologous Pair (Tetrad)

8. Move your chromosomes to the next two circles that are labeled **meiosis I daughter cells.**

✓ Comprehension Check

1. What is the number of chromosomes in each daughter cell? _____

2. Does this number represent the **diploid** or the **haploid** chromosome number?
    _____

3. What happened to the **maternal and paternal** chromosomes from the original parent cell?

    Are the two daughter cells genetically identical? _____ **Explain** your answer.

9. Meiosis continues with a second cellular division.

    Simulate meiosis II **without moving your chromosomes** to another set of circles. Note that the circles you're now using have two labels. In **addition** to meiosis I daughter cells, the circles are also labeled **prophase through telophase of meiosis II.**

    Nothing unusual occurs to the chromosomes in prophase II.

    To simulate **metaphase II,** place the chromosomes **in the correct position** relative to the **equator.**

    To simulate **anaphase II,** separate the two **sister chromatids** by moving them to the opposite poles of the cell.

    Move your chromosomes to the four circles that are labeled **meiosis II daughter cells.** Each of these daughter cells has the potential to develop into a sperm or egg.

Check the movement of your chromosomes through meiosis I and II with your instructor before you continue.

☑ Comprehension Check

1. The original parent cell in your meiosis simulation had two alleles for each trait. How many alleles for each trait are **now** in each daughter cell? _____

2. Are your daughter cells diploid or haploid? _____

3. Are the four daughter cells **genetically identical?** _____ **Explain** your answer.

4. Where in the body does meiosis occur in **males?** _____

   Where does it occur in **females?** _____

Check your answers with your instructor before you continue.

ACTIVITY 3
FERTILIZATION: NATURE'S EQUIVALENT TO ROLLING THE DICE

1. As you know, **only one sperm and one egg** can participate in fertilization.

   Any of the four sperm can potentially fertilize an egg, but the development of female gametes is slightly different. **Only one** of the four daughter cells will develop into an egg cell. The other three aren't functional and are referred to as **polar bodies.**

   You'll simulate fertilization of your gametes with a roll of the dice. Get a **single six-sided die.**

   Are **your** gametes potential sperm or eggs? _____

2. Choose the gamete that will participate in fertilization as follows:

   a. **Number your daughter cells** 1 through 4.

   b. **Roll the die.** If you get a number between 1 and 4, that's your lucky gamete. If you get 5 or 6, roll again.

3. Link up with a group that has produced a gamete of the opposite sex. Take the chromosomes from both gametes and **place them in a circle** labeled **"Fertilized Egg."**

ACTIVITY 4   WHAT'S YOUR BABY'S GENOTYPE?

1. On the two chromosomes in **Figure 17-3,** list your baby's **alleles in their correct order.**

    Solid beads = dominant alleles. Striped beads = recessive alleles.

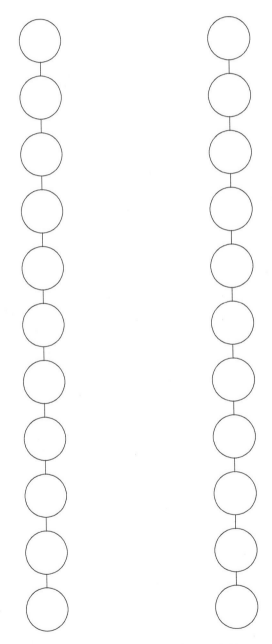

Maternal Chromosome   Paternal Chromosome

FIGURE 17-3. Maternal and Paternal Chromosomes of the Fertilized Egg

ACTIVITY 5   GENOTYPE DETERMINES PHENOTYPE

1. Now that you know your baby's **genotype** for each trait, turn back to **Table 17-1.** In **Table 17-1, highlight or circle** your baby's **genotype and phenotype** for each trait.

    Now that you know your baby's **phenotype** for each trait, you're ready to see what your baby will look like!

2. Choose the appropriate **face shape outline** to begin your baby (**Figures 17-4** and **17-5**).

    Cut out the appropriate **eyes, ears, and other facial features** and use **tape** to attach them to your baby's face (**Figures 17-6 through 17-13**). If your baby has freckles, **draw them** in.

    Add the appropriate **eyelashes** to your baby's face.

    Are you pleased with your results? If not, remember that each time a sperm fertilizes an egg you get a different combination of alleles. You may have better luck next time.

3. **Roll the die again** to determine if your baby is a boy or a girl. If you roll **1 through 3,** it's a **girl.** If you roll **4 through 6,** it's a **boy.**

FIGURE 17-4. Round Face

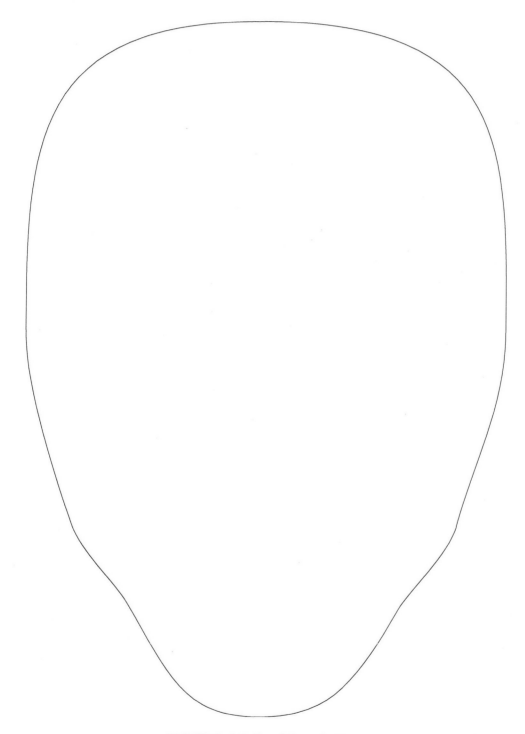

FIGURE 17-5.   Triangular Face

FIGURE 17-6. Curly Hair

FIGURE 17-7. Wavy Hair

FIGURE 17-8. Straight Hair

 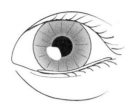

Large Eyes

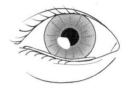 

Medium Eyes

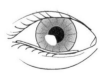 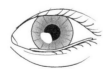

Small Eyes

FIGURE 17-9. Eye Types

Thick Eyebrows   Thin Eyebrows

FIGURE 17-10. Eyebrow Shapes

Thick Lips

Thin Lips

FIGURE 17-11. Lip Types

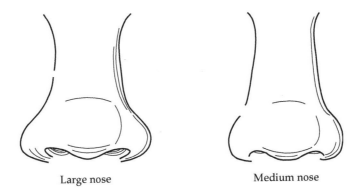

Large nose   Medium nose

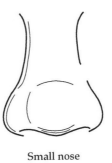

Small nose

FIGURE 17-12.   Nose Size

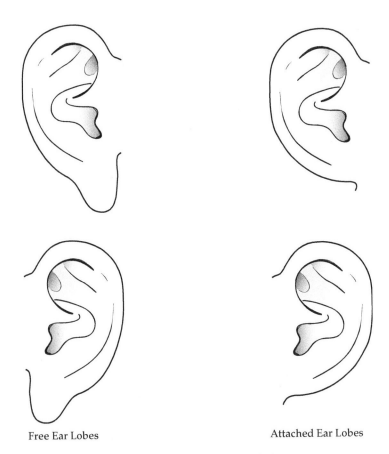

Free Ear Lobes   Attached Ear Lobes

FIGURE 17-13. Ear Lobes

ACTIVITY 6   PASSING ON TRAITS

Twenty-four years have passed. Your baby is now an adult and has married someone who's **heterozygous for freckles.** What is the probability that they'll have a child with freckles?

The first step in calculating the probability of freckles is to **determine the alleles carried by the sperm and eggs** of this couple.

As you know, alleles separate in meiosis, so each sperm and each egg has only **one allele from each homologous pair.**

Your child's genotype for the freckle trait _____ _____

1. A **Punnett Square** is a convenient way to visualize the different combinations of alleles that might occur during fertilization.

2. The spouse's alleles have been entered on the top row of the Punnett Square. Enter the alleles for **your child** along the left side of the square.

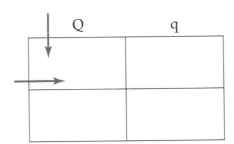

3. Using the Punnett Square as a guide, fill in the boxes by carrying the alleles down from the top and across (from left to right).

   Each box in the Punnett Square represents a **25% probability** of having a baby with that specific genotype.

   What's the probability that this couple will have a child with freckles? _____

ACTIVITY 7   PRACTICING GENETICS PROBLEMS

1. Nicole Johnson is an **albino,** a condition inherited by a **recessive allele (a).** She's marrying a man with normal skin color, whose father was an albino. Nicole makes an appointment for genetic counseling because she wants to evaluate the risk of passing her condition on to her children. What is the probability that a child from this marriage will be an albino? **Show your work.**

2. In humans, assume that **brown eyes (B) are dominant over blue eyes (b).** A blue-eyed man whose father was brown eyed and whose mother was blue eyed marries a brown-eyed woman whose father had brown eyes and whose mother had blue eyes. What are the genotypes of all the individuals? **Show your work.**

    blue-eyed man _____   brown-eyed woman _____

    his father _____   her father _____

    his mother _____   her mother _____

    What is the probability that this couple will have a blue-eyed child? _____

    What is the probability of a brown-eyed child? _____

3. **Polydactyly** (extra fingers or toes) is inherited by a **dominant allele (F).** A father is polydactyl, the mother has the normal number of fingers, and they have had one normal child. What are the genotypes of all the individuals involved? What is the probability that a second child will have a normal number of fingers? **Show your work.**

the father _____   the mother _____

the first child _____

probability of a child with five fingers _____

4. Assume that the baby you made in **Activity 5** has **wavy hair,** inherited through **incomplete dominance (Hh)** and grows up to marry a person who also has **wavy hair.**

What is the probability you'll have grandchildren with **straight** hair? _____

Curly hair? _____

Wavy hair? _____

Show your work.

Self Test

1.   Mark each statement below as **true (T) or false (F).**

_____ Sperm cells have the same number of chromosomes as the man that produced them.

_____ A zygote is a fertilized egg.

_____ Crossing-over occurs in both mitosis and meiosis.

_____ Gametes are genetically identical to each other.

_____ The letters AA are used to indicate a heterozygous phenotype.

_____ Sister chromatids are identical before crossing-over occurs.

_____ Sister chromatids are held together by a structure called synapsis.

_____ Daughter cells produced by meiosis are haploid.

_____ When the chromosomes line up during metaphase I of meiosis, the equator separates the maternal and paternal members of each pair.

_____ The daughter cells produced by meiosis have both a maternal and a paternal chromosome from each homologous pair.

2.   Can a sperm cell contain maternal chromosomes? **Explain** your answer.

3.   Why are chromosomes sometimes drawn as a straight line and other times in an "X" configuration?

4. **Albinism** is a **recessive** condition **(aa)** in which body cells can't manufacture the pigment **melanin,** which colors eyes, hair, and skin. **Normal pigment production is dominant (AA or Aa).** A normally pigmented woman marries a normally pigmented man. To their surprise, they have an albino child. Give the genotypes of the parents and the child:

    the father _____   the mother _____   the child_____

    What is the probability that this couple will have another albino child? _____

    What is the probability that they'll have a normally pigmented child? _____

    Show your work.

5. **Tay-Sachs disease** in humans is controlled by a **recessive allele (t).** This disease is characterized by the inabili ty to produce an enzyme needed to metabolize lipids in brain cells. Without this enzyme, lipids accumulate in the brain cells and gradually destroy the ability of the cells to function. Children affected with this disease usually die by age five.

    What genotypes **must** be found in **both parents** in order to have a child with Tay-Sachs? _____   _____

    Explain your answer and **include a Punnett Square** that shows the possible genotypes among the children of this marriage.

6. Imagine you're a zookeeper in the Big Cat house of the National Zoo in Washington, D.C. Among the tigers, two yellow-coated parents (Ghandi and Sabrina) give birth to a cub named Snowflake that has a rare color variation — a white coat with black stripes.

    Assuming that the **white coat is inherited as a recessive allele (c),** what are the genotypes of the three tigers?

    Ghandi _____   Sabrina _____   Snowflake _____

    If these parents have another cub, what is the probability that the cub will be **heterozygous** for the coat-color trait? **Show your work.**

7. Another zoo wants to start its own tiger exhibit. So far, they have one tiger in their collection, a white one. If the National Zoo sends Ghandi on breeding loan to this zoo, what is the probability that the zoo will get another white tiger for their new exhibit? **Show your work.**

EXERCISE

18

Human Genetics

Objectives

After completing this exercise, you should be able to:

- demonstrate an understanding of the limitations of sample size in scientific data analysis
- determine genotypes using pedigree charts
- explain why more males than females express X-linked traits
- solve genetics problems involving dominant–recessive inheritance, X-linked traits, multiple allele traits, and codominance
- explain the evolutionary relationship between sickle-cell disease and malaria
- apply your knowledge of genetics to real-life situations

CONTENT FOCUS

Not all alleles produce visible traits like skin color or height. Most alleles control **physiological** traits, such as production of digestive enzymes, hormones, and antibodies. Alleles are responsible for invisible traits such as blood type, ability to carry out metabolic pathways (such as producing proteins or storing blood sugar), color vision, and many others. One example of a physiological trait controlled by a **single gene** is the ability to taste a harmless chemical, **PTC** (phenylthiocarbamide).

The story of the discovery of the "bitter taste gene," which controls the ability to taste PTC, is quite interesting. It happened in 1931 at Dupont Chemical Company. Some PTC crystals accidentally blew into the air in the room, and into the mouths of scientists working in that lab. Some of the scientists complained of a bitter taste, but others said the chemical had no taste! After this observation, the scientific method took over. The scientists tested friends, family members, and co-workers, and found similar results. Some tasted PTC as bitter and others couldn't taste it at all. Seventy years later, these results were linked to a single gene located on chromosome #7.

ACTIVITY 1   PTC TASTING

1. Get the following supplies: **one piece of control taste paper and one containing PTC.**

2. **Taste the control paper** first (to establish the taste of the paper itself). Then **taste the PTC paper.**

   If you're a taster, you'll detect a very unpleasant, bitter taste. If there's little difference between the PTC and the control papers, you're not a taster.

   Are you a taster or a nontaster? _taster_

3. Record your taste-test results **on the master chart in the room.**

4. In the general population, approximately 75% of people can taste PTC. The remaining 25% aren't able to taste this chemical.

   Were your class results close to the **expected percentage?** _Yes_

   If not, suggest a possible reason why your class results differed from the expected outcome:

5. The ability to taste PTC comes from a **dominant allele (T).** Using this information, fill in the appropriate genotypes for tasters and nontasters.

   Tasters   _TT_   _Tt_

   Nontasters   _tt_

6. The inheritance pattern of traits like PTC tasting can be diagrammed in a chart called a **pedigree.** A pedigree illustrates the marriages for several generations within a family and the children produced.

   - **Females** are shown with **circles** and **males** are shown with **squares.**

   - A **black square or circle** shows the **presence of the condition** being studied. A **white square or circle** means the **condition is absent** in that person.

   - A marriage or mating is shown by a line connecting the parents.

   - Children from a mating are shown by a vertical line between the parents.

   - All individuals from the **same generation** are shown along the same horizontal line.

7. After examining the **pedigree key,** determine the genotypes of **all the people** in this family. In a few cases, there may not be enough information to determine a person's second allele. In this situation, **enter a question mark (?) in place of the second letter.**

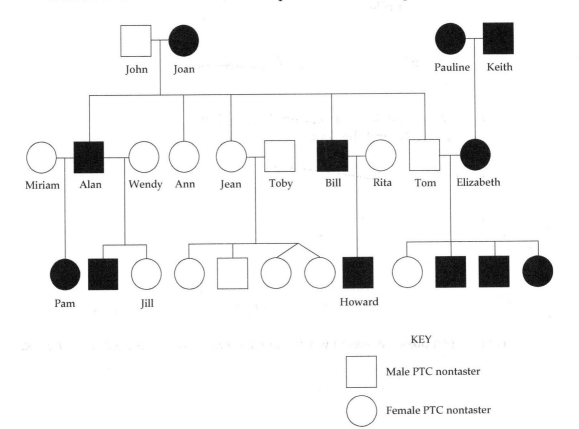

KEY

☐ Male PTC nontaster

◯ Female PTC nontaster

■ Male PTC taster

● Female PTC taster

FIGURE 18-1. PTC Pedigree

8. Referring to the pedigree in **Figure 18-1,** if Jill marries Howard, what is the **probability** that their children will be

tasters? 50% nontasters? 50%

Show your work!

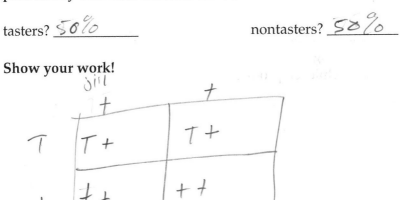

9. If Pam marries Howard, what is the probability that their children will be

    tasters? _75%_   nontasters? _25%_

    Show your work!

    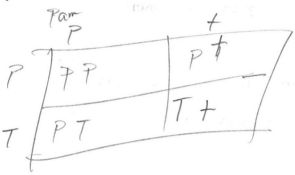

✓ Comprehension Check

1. After performing the PTC experiment, do you think that the ability to taste certain chemicals can explain why your children's food preferences might be different from your own?

    Yes, Having different taste buds changes the flavor of the food. which to Some people will Like it of not Like it.

2. **Barium sulfate** is tasteless to some people, but bitter to others. Two sisters are having their gastrointestinal tracts x-rayed. They are both asked to drink a barium sulfate "milk shake" before the procedure. One sister drinks the "milk shake" without complaint. The other sister complains that it tastes terrible. Using the information gained in today's lab activities, **suggest an explanation for this difference.**

    one is dominte
    one is non dominte.

Check your answers with your instructor before you continue.

ACTIVITY 2 X-LINKED TRAITS

In humans, **the X chromosome is large in comparison to the Y chromosome.** The X chromosome carries information for many traits that aren't related to the sex of the individual. Alleles carried only by the X chromosome are said to be **X-linked** (or sometimes, sex-linked).

Some of the alleles on the tiny **Y chromosome** appear to have no counterparts on X. These **Y-linked alleles** code for traits that are found **only in males.**

Among the X-linked traits are a number of recessive genetic disorders. One of these is **hemophilia,** the inability to produce proteins necessary for blood clotting. Hemophiliacs can bleed to death from relatively minor cuts or bruises. Historical records dating back thousands of years mention the inheritance pattern of hemophilia. Among the ancient Hebrews, sons born to women with a family history of hemophilia were excused from circumcision.

Hemophilia was common during the 1800s in the royal families of Europe, whose members often intermarried. Queen Victoria of England was a **carrier** of the trait. She had one X chromosome with the allele for normal blood clotting (X^H) and the other with the defective allele (X^h). Because she did have **one normal dominant allele,** her blood clotted normally.

Her husband, Prince Albert, was completely normal for this trait. He had one normal allele on the X chromosome (X^H) and a Y chromosome with **no allele related to blood clotting** (Y^o).

Eighteen of Queen Victoria's 69 descendants were carrier females or hemophiliac males. Crown Prince Alexis of Russia was one of these hemophiliac descendants. His affliction indirectly contributed to the overthrow of the monarchy in Russia.

1. Complete this Punnett square for the marriage of Victoria and Albert.

Queen Victoria

Prince Albert	X^H	X^h
X^H	XH XH	XH Xh
Y^o	Yo XH	Yo Xh

2. Considering the **entire Punnett square,** what is the probability that Victoria and Albert could have

   a hemophiliac son __25%__

   a hemophiliac daughter __0__

   a normal son __25__

   a carrier daughter __25__

   a daughter with normal blood clotting __25__

3. A partial pedigree of hemophilia in the descendants of Victoria and Albert is shown in **Figure 18-2. Affected individuals are indicated with darkened circles or squares.** For **each individual in the pedigree,** fill in his/her probable genotype.

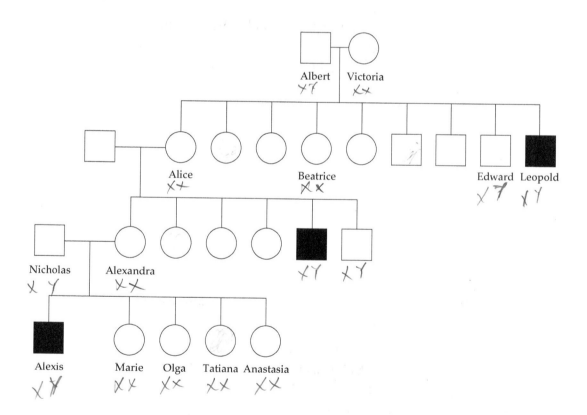

FIGURE 18-2. Partial Pedigree of Hemophilia in the Descendants of Victoria and Albert

✓ Comprehension Check

1. **Challenge Question!** If Victoria and Albert had a hemophiliac daughter, why would it be **unlikely for her to survive puberty** without medical intervention?

once she starts her period she would bleed to death

ACTIVITY 3   CODOMINANCE AND SICKLE-CELL DISEASE

Now that you're beginning to be an expert in solving genetics problems, let's try something a little more challenging. **Sickle-cell disease** is inherited through codominance. **Codominance is a special type of inheritance in which two alleles are equally dominant.**

Both alleles are **expressed independently of each other,** resulting in a **heterozygous individual** that shows **both homozygous phenotypes.** For example, a person with sickle-cell trait has **both** normal and sickled red blood cells in circulation.

Sickle-cell disease is quite common in countries that have a high incidence of malaria. Sickle-cell anemia is a blood disorder affecting transport of oxygen by hemoglobin. The sickled cells block capillaries, depriving the tissues of the oxygen they need.

Individuals with sickle-cell anemia frequently die at an early age. **Heterozygous** individuals have **sickle-cell trait.** These carriers are usually healthy, but experience some problems with intense exercise or under low-oxygen conditions. In the United States, sickle-cell trait affects 1 out of 12 African Americans.

The allele for normal hemoglobin is Hb^A and the sickle-cell allele is Hb^S. The following is a summary of the possible sickle-cell genotypes and phenotypes:

$Hb^A Hb^A$	completely normal
$Hb^A Hb^S$	sickle-cell trait (this person has a combination of normal hemoglobin and the abnormal, sickled form of hemoglobin)
$Hb^S Hb^S$	sickle-cell anemia (all abnormal hemoglobin)

1. If **both parents are heterozygous** for sickle-cell disease, what are the possible genotypes and phenotypes for their children?

	Hb^A	Hb^A
Hb^A	Hb^{A}Hb^S	Hb^{A}Hb^S
Hb^A	Hb^{A}Hb^S	Hb^{A}Hb^S

2. What is the probability of a couple having a child with sickle-cell **trait** if **one** parent is **normal** and the **other** has **sickle-cell trait?**

	Hb^A	Hb^A
Hb^A	Hb^{A}Hb^A	Hb^{A}Hb^A
Hb^S	Hb^{A}Hb^S	Hb^{A}Hb^S

ACTIVITY 4   A FAMILY HISTORY OF SICKLE-CELL DISEASE

The foundation of our modern knowledge of sickle-cell disease is based on the research of Dr. Angela Ferguson and Dr. Roland Scott, professors at Howard University in Washington, D.C. The doctors became interested in sickle-cell disease because many of their friends and some family members were affected by the disease. They published the first research paper on sickle-cell disease in the 1940s — **25 years ahead** of other researchers.

Imagine that you're a physician studying blood genetics in the laboratory of Drs. Ferguson and Scott. The Minister of Health from Nigeria has requested their help to investigate two health problems within the villages of his country that seem to be related: sickle-cell disease and malaria. Drs. Ferguson and Scott send you to investigate.

After spending a year looking into the problem, you've collected data about **Family A** and constructed a **pedigree and a health profile** for this family (see **Figure 18-3**). Each **individual in the pedigree is identified by a number.**

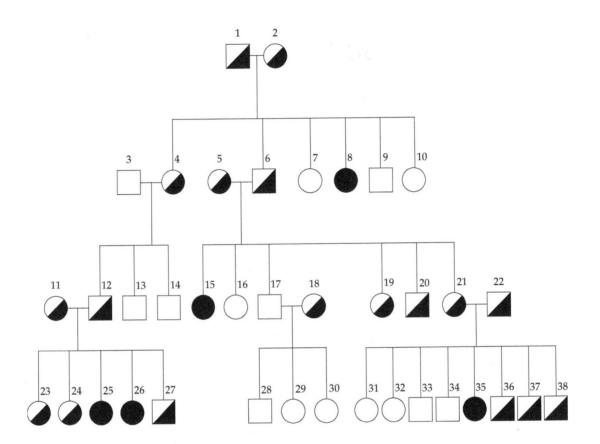

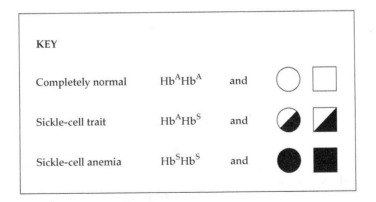

FIGURE 18-3. Pedigree of Family A

Is there any relationship between sickle-cell disease and malaria in Family A? If there is a relationship, why would this occur? To answer these questions, you need some background information about malaria.

Malaria is an infection of the blood carried from person to person by mosquitoes. People of all ages can be infected with malaria, but babies, young children, and pregnant women suffer the worst effects. Malaria affects **10% of the world's population.** It kills about **two million** children every year: that's one every 15 seconds. Most deaths occur in children in sub-Saharan Africa.

Malaria is caused by protozoans (one-celled organisms) of the genus *Plasmodium.*

- Infection begins with a bite from an infected mosquito.

- The parasite travels from the mosquito to your liver, where it begins to reproduce.

- The parasite leaves the liver and travels to the bloodstream, where it infects the **red blood cells.** The parasite reproduces in the red blood cells, which destroys the cells and releases more parasites into the bloodstream.

- The disease causes red blood cells to burst every 48 hours, releasing toxins that cause fever, chills, and even death.

- If another mosquito bites an infected person, that mosquito can then carry the infection to someone else.

As you know, sickled red blood cells have a much lower oxygen-carrying capacity than normal red blood cells. This decreases the ability of the malaria parasites to reproduce.

In addition, the presence of the parasites causes the red blood cells to change into the abnormal, sickled shape. As the blood passes through the spleen, it removes and destroys the abnormal blood cells, and along with it, the malaria parasites. This reduces the severity of the infection.

Therefore, people with sickle-cell trait have a selective advantage for survival in countries where malaria is prevalent.

1. Refer to the **Health Profile of Family A** in **Table 18-1.** On the **pedigree chart in Figure 18-3, place an asterisk** (*) next to the numbers of all persons who are **affected with malaria** or who have **died from malaria.** In **Table 18-2, list the pedigree number** for **each** individual in Family A who has **died from malaria or is currently affected** with this disease.

TABLE 18-1 HEALTH PROFILE OF FAMILY A	
HEALTH STATUS	AFFECTED FAMILY MEMBERS
Affected with malaria	13, 32, 33
Dead from malaria	7, 9, 10, 17, 29, 30, 31, 34
Dead from sickle-cell disease	8, 15, 25, 26, 35
Dead from causes unrelated to malaria or sickle-cell disease	3, 19
In good health	1, 2, 4, 5, 6, 11, 12, 14, 17, 18, 20, 21, 22, 23, 24, 27, 28, 36, 37, 38

TABLE 18-2 GENOTYPES OF INDIVIDUALS AFFECTED BY MALARIA	
LIST THE PEDIGREE NUMBER FOR EACH AFFECTED INDIVIDUAL (SEPARATED BY GENOTYPE)	GENOTYPE
	$Hb^A Hb^A$
	$Hb^A Hb^S$
	$Hb^S Hb^S$

2. **Of the total number of deaths in this family,** what **percentage** died from each cause? Write your answers in **Table 18-3**.

TABLE 18-3 CAUSES OF DEATH IN FAMILY A		
CAUSE OF DEATH	NUMBER DEAD	PERCENTAGE OF TOTAL DEATHS
Malaria		
Sickle-cell disease		
Other		
TOTAL		

3. What is the **leading cause of death** among the members of Family A?

4. **Go back to the pedigree** and examine the genotypes of the individuals who are alive and in good health.

   Among these healthy individuals, what is the most frequent **genotype** for sickle-cell disease? _____

5. Is the death pattern shown by Family A typical of the population as a whole? To answer this question, **use Figure 18-4 to make a graph** comparing:

   a. the percentage of deaths from each cause in Family A (as shown in **Table 18-3**)

   b. the percentage of deaths from each cause in several neighboring communities (as shown in **Table 18-4**)

TABLE 18-4 CAUSES OF DEATH IN NEIGHBORING COMMUNITIES		
CAUSE OF DEATH	NUMBER DEAD	PERCENTAGE OF TOTAL DEATHS
Malaria	1080	
Sickle-cell disease	600	
Other	320	
TOTAL	2000	

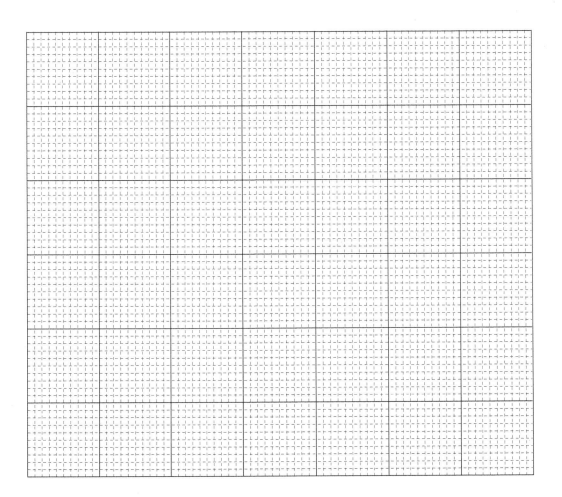

FIGURE 18-4. _____

Check your graph with your instructor before you continue.

✓ Comprehension Check

1. What causes malaria?

2. What causes sickle-cell disease?

3. In a few sentences, explain why the presence of sickle cell trait can protect a person from a malarial infection.

4. In Family A, **family members 21 and 22** are expecting their ninth child. What is the probability that this child will have sickle-cell anemia? **Show your work!**

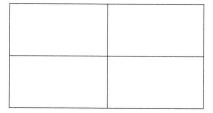

5. Closer to home, Michelle had a brother, Charles, who died of sickle-cell anemia. She is concerned about the chance of the condition appearing in her children. When blood samples were taken and placed under low-oxygen conditions, some of her red blood cells sickled. Those of her husband James, however, remained normal when tested. **Show your work and list the genotypes of all those mentioned in the problem.**

   Michelle _____   Charles _____   James _____

   What is the probability that Michelle's and James's children will have sickle-cell anemia? _____

   Sickle-cell trait? _____

6. **Challenge Question!** Tanya, who has the genotype $Hb^A Hb^S$, and her sister Amelia, who is $Hb^A Hb^A$ are planning a vacation to visit the ruins of a famous Aztec city high in the mountains of Peru. Should either Tanya or Amelia be concerned about taking this trip in view of their sickle cell status? **Explain** your answer.

Check your answers with your instructor before you continue.

ACTIVITY 5   CODOMINANCE AND MULTIPLE ALLELES

Up to this point, we've dealt with traits that have **only two alleles.** Human blood types (A, B, and O) are inherited by **multiple alleles.** There are three possible alleles for blood type. **Two of the three** possible alleles are **codominant.**

Codominance is a special type of inheritance in which two alleles are equally dominant. Both alleles are expressed independently of each other, resulting in a heterozygous individual that shows both homozgous phenotypes. For example, red blood cells with both A and B proteins in their cell membranes.

Multiple alleles means that there are **more than two possible alleles** for that trait, although each person inherits only two of the three possible alleles (one from their mother and one from their father).

In human blood types, the alleles I^A and I^B are codominant. Both I^A and I^B are **dominant** over the **recessive allele i.**

GENOTYPE	PHENOTYPE
$I^A I^A$ or $I^A i$	Type A blood
$I^B I^B$ or $I^B i$	Type B blood
$I^A I^B$	Type AB blood
ii	Type O blood

1. Howard Hughes was an inventor, multimillionaire businessman, and adventurer, whose life was the subject of the 2004 film "Aviator." When he died, he left no legitimate heirs. Soon, however, a long succession of people claiming to be his children began to appear.

   A young man claiming to be Howard Hughes's child sued for a share of the estate. The judge ordered blood tests to determine the validity of the claim. Howard Hughes had blood **type AB,** the **mother** of the young man had blood **type A,** and the **young man** himself had **type O** blood. If you were the judge, how would you rule? **Explain** your answer.

2. Is it possible for a **type A** person married to a **type B** person to have **type O** children? **Explain** your answer.

Check your answers with your instructor before you continue.

Self Test

1. The ability to taste PTC is due to a **dominant allele (T).** A woman nontaster married a man who was a taster. They had three children. Their two sons were tasters, but their daughter was a nontaster. All four grandparents were tasters. **What are the genotypes** of all the individuals mentioned?

   The woman _____   Her husband _____

   The two sons _____   The daughter _____

   Grandparents _____

2. Your sister died from Tay-Sachs disease, inherited as a **recessive** allele **(t).** You're married and planning to start your family. You're worried about the disease and decide to have genetic testing to see if you or your spouse is a carrier of the Tay-Sachs allele. The test results show that you're a carrier of the allele, but your spouse isn't.

   What is the probability that you and your spouse will have a child with Tay-Sachs disease? **Show your work.**

3. Red–green color blindness is inherited through an **X-linked, recessive** allele **(b).** Two parents, Fred and Ginger, have normal vision. They have two daughters, Takiyah and Kelly, who also have normal vision, and a color-blind son, David.

   Daughter Kelly has a color-blind son, Kevin. Daughter Takiyah has five sons, all with normal vision. What are the genotypes of all the individuals? **Show all your work!**

   Fred _____   Ginger _____   David _____

   Takiyah _____   Kelly _____   Kevin _____

   Takiyah's five sons _____

   If Kelly marries a man with normal vision, what is the probability that she'll have

   a color-blind son? _____   a color-blind daughter? _____

4. Ralph has normal blood clotting, but he has two brothers and a sister who have hemophilia (an **X-linked, recessive** disorder). What are the most probable genotypes of Ralph's parents? **Explain** your answer.

5. Albinism in humans is expressed as the **absence** of pigment from the skin, hair, and eyes. Using the information in **Figure 18-5**, determine whether albinism is inherited as a **dominant or as a recessive** trait. **Affected** individuals are represented by **shaded** squares and circles.

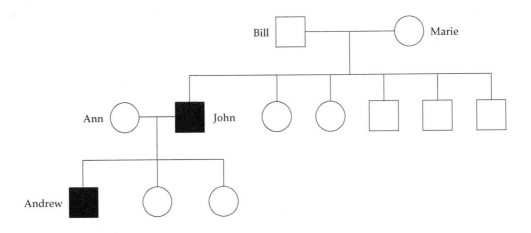

FIGURE 18-5.   Pedigree Showing Albino Individuals

a. Underneath **each circle and square** in the pedigree, **enter the genotype** of that individual.

b. **(Circle one answer).** Albinism is a trait that is probably inherited through a **dominant / recessive allele.**

Support your answer with information from the pedigree.

6. Neil and Alice are concerned because their newborn daughter doesn't appear to resemble either of them. They suspect there was a mix-up at the hospital. They check the blood type of the baby and find it is **type O.** Because Neil has **type A** blood and Alice has **type B,** they conclude that a mistake has been made. Are they correct? **Explain** your answer.

Introduction to Molecular Genetics

Objectives

After completing this exercise, you should be able to:

- demonstrate an understanding of DNA structure and the base-pairing rule
- explain the roles of the following in protein synthesis: DNA, mRNA, tRNA, and ribosomes
- complete the processes of transcription and translation from a strand of DNA and determine the sequence of amino acids in the resulting polypeptide chain
- explain how changes in the DNA code may affect protein synthesis and cause health problems

CONTENT FOCUS

You've just spent some time thinking about genetics and the different ways in which traits can be transmitted from parents to children. Each chromosome is divided into sections called **genes** that are the basis of inheritance. The traits you inherited through your genes (such as hair color, blood type, presence of sickle-cell anemia, or ability to produce the hormone insulin) are all controlled by the production of proteins. **Genes contain the coded instructions your body uses to assemble the hundreds of different types of proteins that make you a unique individual.**

Genes are composed of a molecule known as **deoxyribonucleic acid (DNA)**. In this exercise, you'll take a closer look at the **DNA molecule** and the role it plays in all the cells of your body.

ACTIVITY 1   REMOVING DNA FROM CELLS

> ### Note:
> Read these directions COMPLETELY before proceeding.

1. Work in groups. Get the following supplies: **one piece of calf thymus (5 grams), one graduated 10-ml pipette with manual dispenser, one 50-ml test tube, a test tube rack, a small bowl, and a pair of scissors.**

2. On your laboratory table, you'll find a **blender, a piece of cheesecloth, a 500-ml graduated cylinder, a container of 6% saline solution, rubber bands, and a 600-ml beaker.**

    Using **scissors, mince** the piece of thymus **as much as possible** and place the pieces in the bowl.

    Add **100 ml of 6% saline solution** to the bowl.

> ### Note:
> All groups at your table will be using the blender at the same time, so you MUST coordinate your activities!

3. Coordinating with your partner groups, pour the saline containing the **minced thymus** from **each of the groups** at your laboratory table into the blender.

4. **Put the lid on the blender.** Blend the thymus mixture **on a low setting** for several minutes until no large pieces remain.

5. **Remove the lid** from the blender.

    Assemble **eight thicknesses of cheesecloth** (four pieces from the package, which is manufactured in double thickness). Place the pieces of cheesecloth over the top of the blender and **secure** the cloth **tightly** with the rubber band.

6. Making sure the cheesecloth is **tightly** fastened on the top of the blender, **pour the contents (filtered extract)** into the **600-ml beaker.**

> ### Note:
> At this point, you'll go back into your ORIGINAL groups and complete the experiment.

7. Using a **graduated 10-ml pipette with manual dispenser,** remove **10 ml of filtered extract** from the beaker and place the extract into a **50-ml test tube.** Place the test tube in the rack.

   Place your **used pipettes** in the waste container.

8. Get **one graduated 2-ml pipette with manual dispenser, one graduated 10-ml pipette, and a thin glass rod.**

   In your supply area, you'll find a container of **10% SDS solution** (sodium dodecyl sulfate) and **an ice bucket containing 15-ml screw-cap tubes of 95% ethanol.**

9. To the **test tube containing the filtered thymus** extract, add **1 ml of SDS solution.** Tap the test tube several times to **gently** mix the contents.

Note:
Read the following directions COMPLETELY before proceeding.

---

10. Select a **tube of ethanol** from the ice bucket. Using a **clean,** graduated 10-ml pipette, add **10 ml of ice-cold 95% ethanol** to the test tube.

    With the test tube in the rack, place the filled pipette with its tip against the **inside wall** of the test tube. SLOWLY allow the ethanol to dribble down the inside of the tube, as **demonstrated in Figure 19-1.** Don't shake the test tube during this procedure.

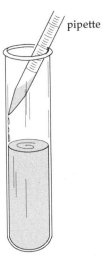

pipette

FIGURE 19-1. Method for Adding Ethanol to Filtered Thymus Extract

11. The ethanol is lighter than the contents of the tube. When added according to directions, the ethanol will form a **clear layer ABOVE** the filtered thymus extract (see **Figure 19.2**).

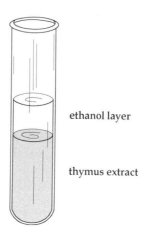

ethanol layer

thymus extract

FIGURE 19-2.   Ethanol Layered Above Filtered Thymus Extract

12. Observe the test tube for **five minutes.** The DNA will gradually separate from the thymus mixture and rise into the ethanol layer.

    Describe the appearance of the DNA.

    Looks like a cloady supstance

13. Let the DNA sample remain undisturbed while you complete the following **Comprehension Check** and **Activity 2.** When you've completed Activity 2, **observe** your DNA sample and see if any changes have occurred. **Record your observations** below.

14. To remove the accumulated DNA from the test tube, follow the directions for **DNA spooling** below:

    a. **Gently insert** the glass rod through **the ethanol layer** into the **accumulated DNA.**

    b. **Carefully twirl** the rod between your fingers, winding the DNA onto the rod, imitating thread on a spool.

    c. **Slowly remove the rod** from the test tube.

✓ Comprehension Check

1. The thymus is an organ made up of many different types of cells. What percentage of thymus cells contain DNA?

   a. 0%   d. 75%
   b. 25%   e. 100%
   c. 50%

   Explain your answer.

2. If you removed DNA from a sample of human thymus, how many separate DNA molecules would be removed from **each cell?** _____

3. **(Circle one answer.)** If you repeated the same experiment with an equal number of kidney cells, the amount of DNA collected would **increase / decrease / stay the same.**

   Explain your answer.

4. **(Circle one answer.)** If you repeated the same experiment with an equal number of sperm cells, the amount of DNA collected would **increase / decrease / stay the same.**

   Explain your answer.

5. **Challenge Question! (Circle one answer.)** If you repeated the same experiment with an equal number of red blood cells, the amount of DNA collected would **increase / decrease / stay the same.**

   Explain your answer.

Check your answers with your instructor before you continue.

ACTIVITY 2   THE BASICS OF DNA STRUCTURE

The extraction of DNA that you just performed showed that DNA is present in cells. It does not, however, give you much information about its actual structure.

DNA molecules are composed of small building blocks called **nucleotides.**

1.   Each DNA nucleotide is composed of three smaller molecules hooked together:

     one five-carbon sugar (deoxyribose)

     one phosphate

     one nitrogen base

2.   Four different types of nucleotides are needed to build a DNA molecule.

     Each of these four nucleotides has a **different nitrogen base: adenine, guanine, cytosine, or thymine.**

3.   DNA has a structure similar to a ladder. The two sides of the ladder are composed of **alternating sugar and phosphate molecules.**

4.   Each sugar molecule is attached to **one nitrogen base.** The two strands of DNA are attached by bonds between the nitrogen bases on each side of the ladder.

     Each nucleotide base only bonds with **one specific partner.** The combination of two bases is called a **base pair.**

     Adenine always bonds with thymine. A-T

     Guanine always bonds with cytosine. G-C

5.   Fill in the blanks on the incomplete DNA molecule in **Figure 19-3.**

     Use the symbol **"P"** for **phosphate,** **"S"** for **sugar,** and **"A," "T," "C," or "G"** for the **appropriate nitrogen bases.**

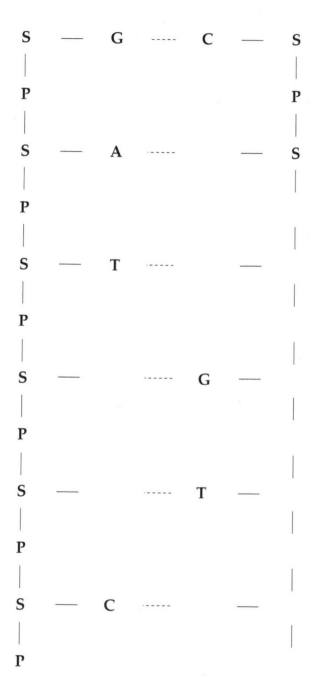

FIGURE 19-3. Incomplete DNA Molecule

✓ Comprehension Check

1. On **Figure 19-3,** draw a box around **one complete nucleotide.**

2. **How many nucleotides** are shown in the DNA molecule in **Figure 19-3?** _____

3. How many **different** types of nucleotides were used to construct the DNA molecule in **Figure 19-3?** _____

Check your answers with your instructor before you continue.

ACTIVITY 3   BUILDING A MODEL OF DNA

1. Work in groups. Get a **DNA model kit.**

2. Construct the **support stand for the model** by assembling:

   one gray tube (8″ long)

   three green tubes (2″ long)

   one black connector with four prongs

   Set the stand aside for later use.

3. Separate the parts of the DNA model according to the descriptions in **Table 19-1.**

TABLE 19-1 PARTS OF THE DNA MODEL	
PARTS OF THE DNA MOLECULE	DESCRIPTION OF MODEL PART
Deoxyribose sugar	Black, three prongs
Phosphate	Red, two prongs
Sugar-to-phosphate connectors	Yellow tube
Adenine (A) base	Blue tube
Thymine (T) base	Red tube
Guanine (G) base	Green tube
Cytosine (C) base	Gray tube
Base-to-base connectors	White, two prongs

4. Using your knowledge of DNA structure, **assemble the DNA model.** Use the same **base-pairing sequence** as that shown in **Figure 19-3.**

5. Place the model onto the stand by inserting the vertical tube on the stand through the holes in the **white base-to-base** connectors.

    Twist the model as you slide it onto the stand. Stop twisting when the entire molecule fits onto the stand.

    The twisted shape of the DNA molecule is known as a **double helix.**

✓ Comprehension Check

1. How many nucleotides are present in your DNA model? _____

2. What is the base-pairing rule?

3. Why would it be appropriate to call a DNA molecule a **polynucleotide?**

4. Assume that the model you just built is an exact representation of **your** DNA code.

    a. Would you use the same bases to construct your lab partner's DNA? _____

    b. Would you assemble the bases in the same order to make a model of your lab partner's DNA? _____

    Explain your answer.

Check your answers with your instructor before you continue.

Genes have the instructions to make polypeptide chains. These instructions are part of the genetic code. Polypeptide chains are the structural units of proteins. A polypeptide chain is made of many amino acids hooked together.

The key to the genetic code is the sequence of nitrogen bases along **one side** of the DNA molecule. To construct a protein, you must know the **order of the bases.** The code is written in **three-letter "words."** Each of these words (called **triplets**) tells the cell which amino acid should come next when building a protein.

For proteins to function properly, the amino acids must be assembled in the correct order.

5. How many **triplets** are present along one side of your DNA model? _____

6. How many **amino acids** will be present in the protein made from your model? _____

Note:
DON'T take your DNA model apart yet.

ACTIVITY 4   STEPS OF PROTEIN SYNTHESIS

When a particular protein is needed by the body, regions of the double helix unwind so that a cell gains access to the genes that contain the coded information to make that protein. Protein synthesis has two steps: **transcription** takes place in the nucleus and **translation** occurs in the cytoplasm. Both steps require molecules of **RNA (ribonucleic acid).**

Although the **nucleus** contains the **instructions** for protein synthesis, the **machinery to make proteins** is located in the **cytoplasm.** The coded information is **transferred** from the nucleus to the cytoplasm during **transcription.**

Genes begin with an **initiator (start) codon** (AUG) that also codes for the amino acid methionine. In addition, the end of each mRNA sequence has a **terminator (stop) codon** that signals the ribosome that it is at the end of the amino acid sequence.

For example, consider the mRNA transcription of the following DNA sequence:

DNA: T A C   A G A T A A C C G C G   A C T

mRNA: A U G   U C U A U U G G G C G C   U G A

          start   coding   stop
          codon   region   codon

Transcription

1. During transcription, DNA bases are **copied** to form a single strand of RNA, called **messenger RNA (mRNA).** As with the DNA, mRNA is divided into coded three-letter words. In mRNA, these words are called **codons.**

2. The **base-pairing rule** is used to form messenger RNA **with one exception.** RNA molecules **don't** have the nitrogen base **thymine.** They have uracil instead.

    Base-pairing to form mRNA:

DNA BASE	mRNA BASES
C	G
G	C
T	A
A	U

3. The coded information to make a protein appears along **one side** of the double helix. Practice transcription by filling in the correct messenger RNA codons in **Table 19-2.**

T A B L E 1 9 - 2 TRANSCRIPTION OF mRNA								
DNA TRIPLETS	CGC	ATA	GAC	TTT	CTT	ACT	TAG	CAT	AAA
mRNA codons									

Note:

To make it easier for you to practice transcription and translation, the start and stop codons have been removed from all the DNA sequences and only the coding regions of the DNA are shown.

---

4. How many **amino acids** would be in this protein? _____

5. Which of the five types of nitrogen bases is **not** found in mRNA? _____

Translation

1. A cell needs **amino acids** to construct proteins. The amino acids are carried to the ribosomes by another type of RNA molecule, called **transfer RNA (tRNA).**

   A tRNA molecule has **two functional ends.**

   One end picks up amino acids in the cytoplasm (see **Figure 19-4**).

2. The other end is called the **anticodon.** It contains **three nitrogen bases** that can form a base pair with a **matching codon** in the messenger RNA.

3. Each type of tRNA can carry **only one type of amino acid.**

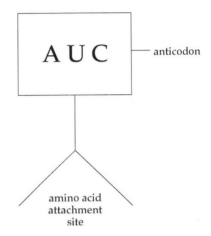

FIGURE 19-4. Structure of a Transfer RNA Molecule

There are enough different types of transfer RNA molecules to carry all the different types of amino acids needed to make your body's proteins.

4. **Where do the transfer RNA molecules take the amino acids?**

   They take them to **ribosomes,** organelles in the cytoplasm where proteins are manufactured.

   Ribosomes are made of proteins and a third type of RNA called **ribosomal RNA.**

5. **Ribosomes read messenger RNA codons and accept amino acids brought by transfer RNA molecules.**

   Ribosomes hook amino acids together in the **order specified by the messenger RNA** codons to construct the polypeptide chain.

Summary of Protein Synthesis

■ DNA contains the instructions to make polypeptide chains. A region of the DNA double helix unwinds. Coded instructions to make a protein are exposed.

■ This information is carried to the cytoplasm by messenger RNA molecules (transcription).

■ Amino acids in the cytoplasm are used to build polypeptides.

■ Transfer RNA molecules pick up the amino acids and transport them to ribosomes, the locations where proteins are made.

■ Ribosomes bond amino acids together according to the instructions in the genetic code.

ACTIVITY 5   BUILDING A REAL PROTEIN

Imagine the following situation: you're about to give birth to a baby. The brain produces the hormone **oxytocin** (a small protein), which causes uterine muscles to contract for childbirth. After birth, this same hormone causes muscles in the mammary glands to contract, releasing milk for nursing the baby.

Suppose this is your first baby. How does the brain know how to manufacture oxytocin if it has never been needed before? The information is stored in your DNA "reference" library.

1. Using the steps outlined in **Activity 4** as guidelines, build the protein **oxytocin.**

    Oxytocin is one of the smallest proteins with only nine amino acids connected into a single polypeptide chain.

2. **Transcribe** the DNA triplets that code for oxytocin into **messenger RNA codons** and add them to the appropriate spaces in **Table 19-3.**

TABLE 19-3 TRANSCRIPTION OF OXYTOCIN									
DNA TRIPLETS	ACG	ATG	TAT	GTT	TTG	ACG	GGA	GAC	CCC
mRNA codons									

3. The messenger RNA must detach from the DNA and leave the nucleus for **translation** to take place.

   Cut out the strip of mRNA and move it to the ribosome in Figure 19-5. Notice that the ribosome has two parts, separated by a groove. The mRNA should be placed along the groove between the upper and the lower sections.

4. Each transfer RNA molecule in **Figure 19-6** is ready to deliver an amino acid to the ribosome.

   Where should each transfer RNA molecule deliver its amino acid?

   Place each tRNA molecule under the correct mRNA codon.

| **Hint:** |

You'll be able to match each tRNA with its codon by following the base-pairing rule.

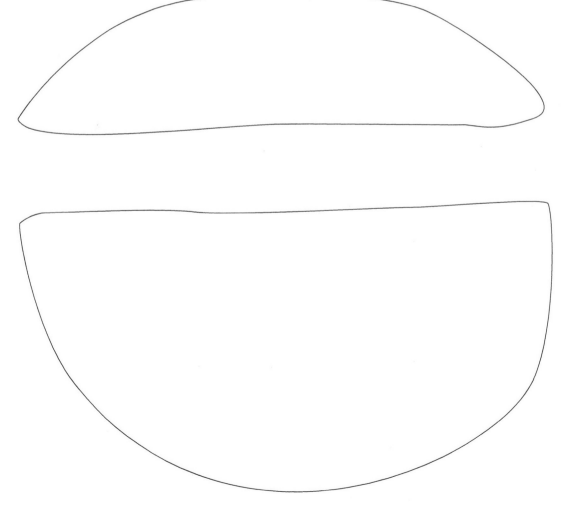

FIGURE 19-5. Translation

To determine the **sequence of amino acids in the protein oxytocin**, match the **shape** of each amino acid **symbol** (attached to the bottom of the tRNA molecules in **Figure 19-6**) to the **amino acid names** in **Table 19-4.** To do this, read the tRNA sequence from left to right.

Fill in the blanks below with the names of the amino acids.

Oxytocin amino acid sequence:

Amino acid #1   _____

Amino acid #2   _____

Amino acid #3   _____

Amino acid #4   _____

Amino acid #5   _____

Amino acid #6   _____

Amino acid #7   _____

Amino acid #8   _____

Amino acid #9   _____

Check your oxytocin molecule with your instructor before you continue.

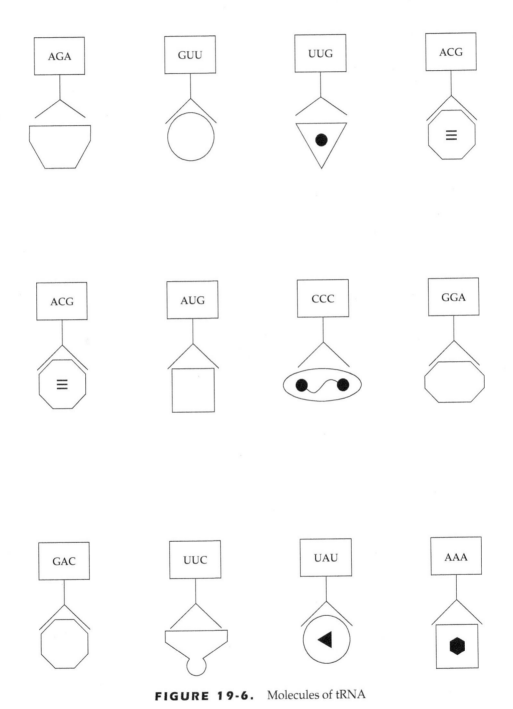

FIGURE 19-6. Molecules of tRNA

T A B L E   1 9 - 4

KEY TO IDENTIFYING AMINO ACIDS

SHAPE	NAME OF AMINO ACID
	histidine
	glutamine
	asparagine
	cysteine
	tyrosine
	glycine
	proline
	leucine
	alanine
	isoleucine
	serine

✓ Comprehension Check

1. If the second amino acid in your polypeptide chain were changed to **valine,** would the protein still be oxytocin? _____ **Explain** your answer.

2. **(Circle one answer.)** The DNA triplet A A A would be transcribed into the mRNA codon **T T T / U U U.**

3. Arrange the following steps of protein synthesis in the correct order.

   _____ tRNA molecules pick up amino acids

   _____ mRNA transcribed

   _____ DNA double helix unwinds

   _____ mRNA binds to ribosome

   _____ Ribosome bonds amino acids together

   _____ tRNA anticodon links with mRNA codon

   _____ mRNA leaves nucleus

   _____ Polypeptide chain completed

4. If you include the start and stop codons that were removed from the mRNA sequence for oxytocin, how many DNA triplets were in the **original gene** that coded for oxytocin? _____

   Based on your answer to the question above, how many **additional nucleotide bases** would have been present in the **original transcribed mRNA** strand for oxytocin? _____

Check your answers with your instructor before you continue.

Self Test

Matching Answers can be used **more than once.**

    a. DNA   c. both DNA and RNA
    b. RNA   d. neither DNA nor RNA

1. _____ Present in chromosomes.

2. _____ Contains the base adenine (A).

3. _____ Structure is a double helix.

4. _____ Contains the base thymine (T).

5. _____ Is made of amino acids linked together.

6. _____ Part of the structure of ribosomes.

7. _____ Made of nucleotides linked together.

8. _____ Contains the base uracil (U).

9. _____ Contains the sugar deoxyribose.

10. _____ Contains phosphate.

11. A friend of yours has volunteered for a study on a new type of gene therapy. He tells you that the researchers intend to examine his circulating red blood cells to determine whether the gene was successfully inserted into a chromosome. Has your friend misunderstood the planned procedures? **Explain your reasoning.**

We have a "mini-gene" with the sequence **T A T C C T G A T T C A A A A G T T**.
Given this information, answer the following questions:

12. Assuming that there are no start or stop triplets present, how many amino acids would be in the protein encoded by this gene? _____

13. The last codon in the mRNA will be _____ .

14. The anticodon of the t-RNA which carries the **LAST** amino acid will be
_____.

15. Using **Table 19-5**, list the **FIRST FOUR** amino acids of this protein in their **correct sequence.**

    Amino acid #1   _____

    Amino acid #2   _____

    Amino acid #3   _____

    Amino acid #4   _____

16. If a mutation changed the DNA code of the second triplet from **C C T to C C C,** would that change the amino acid structure of the protein made using this code? **Explain** your answer.

TABLE 19-5

MESSENGER RNA CODONS AND THEIR CORRESPONDING AMINO ACIDS

1ST LETTER	2ND LETTER	3RD LETTER	AMINO ACID	1ST LETTER	2ND LETTER	3RD LETTER	AMINO ACID
A	A	A	lysine	U	A	A	terminator 1
		C	asparagine			C	tyrosine
		U	asparagine			U	tyrosine
		G	lysine			G	terminator 2
	C	A	threonine		C	A	serine
		C	threonine			C	serine
		U	threonine			U	serine
		G	threonine			G	serine
	G	A	arginine		G	A	terminator 3
		C	serine			C	cysteine
		U	serine			U	cysteine
		G	arginine			G	tryptophan
	U	A	isoleucine		U	A	leucine
		C	isoleucine			C	phenylalanine
		U	isoleucine			U	leucine
		G	methionine/start			G	leucine
C	A	A	glutamine	G	A	A	glutamic acid
		C	histidine			C	aspartic acid
		U	histidine			U	aspartic acid
		G	glutamine			G	glutamic acid
	C	A	proline		C	A	alanine
		C	proline			C	alanine
		U	proline			U	alanine
		G	proline			G	alanine
	G	A	arginine		G	A	glycine
		C	arginine			C	glycine
		U	arginine			U	glycine
		G	arginine			G	glycine
	U	A	leucine		U	A	valine
		C	leucine			C	valine
		U	leucine			U	valine
		G	leucine			G	valine

Biotechnology:
DNA Analysis

Objectives

After completing this exercise, you should be able to:

- discuss the differences between genes and noncoding regions of DNA
- summarize how the process of gel electrophoresis separates DNA molecules
- explain the role of noncoding DNA regions in producing a DNA profile
- draw conclusions that are supported by analysis of RFLP and STR profiles
- interpret DNA data presented in the form of tables, charts, and/or graphs
- explain how STR analysis can be used to determine paternity
- apply your knowledge of DNA fingerprinting to real-life situations

CONTENT FOCUS

Half of your 46 chromosomes were inherited from your mother and the other half from your father. Consequently, **everyone has a unique DNA pattern** (except for identical twins). Tissue samples containing DNA can be used for identification in criminal cases, in paternity suits, and in cases where visual identification is not possible. The technique used to make these genetic comparisons produces a **DNA profile (also called a DNA fingerprint).**

DNA for examination is isolated with a process similar to the method you used in the DNA spooling laboratory. As you know, DNA is located in the nucleus of all body cells. Therefore, DNA can be extracted from any tissue. Common sources include **white blood cells, hair roots, semen, saliva** (which contains **epithelial cells**), and other body tissues.

So far, you've studied basic genetics and the structure of the DNA molecule. You've seen that various traits are produced on the basis of your genetic code (DNA). However, not all of a person's DNA codes for synthesis of proteins. **DNA molecules also have areas that are not genes.**

Originally, this noncoding DNA was called **"junk DNA"** because scientists didn't understand or recognize the function for these sections of the chromosomes (see **Figure 20-1**). They originally believed that these sequences were present to simply fill the gaps between the genes. Although the functions for most noncoding regions have not yet been discovered, scientists are beginning to find evidence that some of these repetitive DNA sequences play important roles in cellular metabolism and inherited diseases.

Gene 1   Noncoding DNA   Gene 2

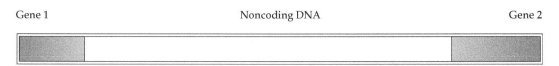

FIGURE 20-1.   Position of Genes and Noncoding Regions of DNA

Regions of noncoding DNA vary a great deal among individuals, and when specific DNA regions are studied, scientists can use this information to establish human identity, analyze evolutionary trends, and determine predispositions to certain diseases.

The pattern of noncoding regions between any given two genes is unique and always has the same repeating pattern of nucleotides (for example **C A T**). The number of times these nucleotides are repeated, however, is highly variable among people.

As you can see in **Figure 20-2,** Person A has **nine** repeating **C A T** segments, while Person B has only **five.** Some people have dozens of repeats of the same pattern.

Each person has a distinct **DNA profile** (or fingerprint).

Person A

Person B

FIGURE 20-2.   Repeating Noncoding DNA Segments

One of the original processes used to create a DNA profile is called **RFLP Analysis.** RFLP stands for **restriction fragment length polymorphism.**

Fragment length polymorphism refers to the fact that the **pieces (fragments) of noncoding regions are different lengths in different people.**

To make a reliable "RFLP match," molecular geneticists use several different non-coding DNA regions. When several different regions are compared, the odds against a mistaken match can be **more than a billion to one.**

ACTIVITY 1   RFLP FINGERPRINTS

An early form of forensic DNA analysis used **restriction enzymes (enzymes that cut the DNA molecule at specific sites)** to produce fragments of DNA. These fragments vary in size from individual to individual based on the number of times that a core sequence is repeated.

Fragments of DNA are separated using a process called **gel electrophoresis.** The **gel** is a thin sheet of gelatin supported by a glass plate with **electrodes** attached at both ends. One end of the gel has a **positive** charge and the other end has a **negative** charge.

The DNA fragments are attracted to the **positive end** of the gel plate. **Smaller, lighter** fragments migrate more easily through the gel and therefore travel farther in a given time period than larger, **heavier** fragments. The light and heavy fragments form bands across the gel layer (see **Figure 20-3**).

The DNA bands of interest are **marked** with special **molecular probes.** Probes are small pieces of DNA that use the **base-pairing rule** to locate and bind **only** to the fragments that will be used to form the DNA profile.

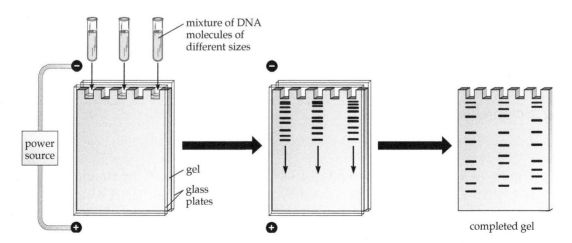

FIGURE 20-3. Gel Electrophoresis of DNA

The sheet of DNA bands is "photographed" with **X-ray film.** The **only** bands visible on the developed film are those that have been **labeled with molecular probes.** The X-ray **pattern of bands** for several DNA locations is analyzed to create a **DNA profile** for an individual (see **Figure 20-4,** which contains samples of several DNA profiles). Each profile is represented by a **vertical column of DNA bands.**

For RFLP testing, **several different regions (called loci) of non-coding DNA are analyzed.** If **ANY** of the loci **don't match** between an evidence sample and a known individual, **the person is eliminated** as a possible source of the evidence DNA.

If, however, **all of the loci in the evidence sample** exhibit the **same pattern as the known** sample, calculations are conducted to determine how rare or common the profile is.

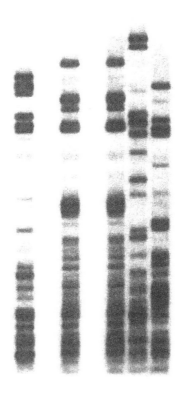

FIGURE 20-4.   Sample DNA Fingerprints

✓ Comprehension Check

Figure 20-5 shows some samples of DNA profiles produced by multiple tests of RFLP loci. Each profile is represented by a **vertical column of DNA bands.**

Two profiles match and two do **not** match. Matching profiles must have **exactly** the same banding patterns. The bands have been numbered for reference.

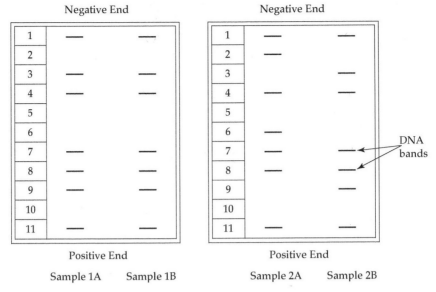

FIGURE 20-5. Four DNA Profiles

1. **(Circle one answer.)** The DNA sample was placed in the gel at the **negative /** **positive** end of the tray.

2. List the bands that **don't match** in samples 2A and 2B. 2, 3, 5, 6, 9, 10

3. Which band in each sample contains the **heaviest** DNA fragments? 2A

4. Which band contains the **lightest** fragments? 1A

5. How many DNA bands are present in **sample 2B?** 7

6. **(Circle one answer.)** If the DNA fragments are migrating toward the positive end of the gel, DNA molecules must have a **negative / positive** charge.

7. **Challenge Question!** What can you conclude from the RFLP analysis if you know that the DNA in **sample 1A** is from blood and the DNA in **sample 1B** is from skin?

   They are from the same person.

Check your answers with your instructor before you continue.

ACTIVITY 2   SOLVING A CRIME USING RFLP FINGERPRINTS

On January 1, at about 2:00 A.M., the police responded to a report of gunshots at the 600-block of Knorr Street. Arriving at the scene, the officers found a parked Dodge pickup. In the front seat was a deceased male with several gunshot wounds.

Shoe marks were visible in a pool of blood outside the truck, but they weren't clear enough to identify the type of footwear that made them. Homicide detectives have narrowed the field to three suspects.

Suspect A was arrested near the scene. He was wearing black work boots that appeared to have dried blood on the soles.

At the home of **Suspect B,** investigators recovered a pair of tennis shoes that also appeared to have dried blood on the soles. Suspect B says she had a severe nosebleed, which might have resulted in blood on her shoes.

Suspect C was the roommate of the deceased. When officers went to the apartment to notify him of the death, Suspect C was packing a suitcase. He was wearing high-top sneakers that appeared to have dried blood on the soles and sides. When questioned about the blood and the murder, Suspect C had no comment.

All the samples were tested and identified as human blood. The Homicide detectives have asked the Central Police DNA Laboratory to analyze the blood evidence using RFLP analysis. You are the forensic scientist assigned to the case. **Figure 20-6** shows the DNA profiles from all the blood samples.

1. Note that the box under the **first column** in **Figure 20-6** (blood of victim) contains the number **1.** This is the first unique DNA profile found.

2. Moving from left to right, examine each DNA profile.

    If the new profile **doesn't match any previously examined profile,** enter a **new number** in the box under the column (**2, 3, 4,** etc.).

    If the new DNA profile **is an exact match** of a previous profile, enter the **number you gave to the profile that matches this new profile.**

    Continue with this process until you've examined all the DNA profiles in **Figure 20-6.**

Note:
When trying to determine whether two DNA profiles are a match, remember that 100% **of the bands have to be the same. If even one band doesn't match, the profiles are not from the same person.**

	Blood of Victim	Blood on Ground at Crime Scene	Blood of Suspect A	Blood of Suspect B	Blood of Suspect C	Blood from Suspect A's Shoes	Blood from Suspect B's Shoes	Blood from Suspect C's Shoes
1			—		—	—		
2	—	—						—
3	—	—			—			—
4			—			—		
5			—	—		—	—	
6				—	—		—	
7	—	—		—			—	—
8					—			
9			—		—	—		
10	—	—						—
11				—			—	
12			—		—	—		
13				—			—	
14								
15	—	—						—
16	—	—						—
17								
18				—			—	
19								
20	—	—	—			—		—
	1	1	2	3	4	2	3	1

FIGURE 20-6. DNA Profiles — Knorr Street Homicide

✓ Comprehension Check

1. How many **unique** profiles were found? _____4_____

2. Did any of the profiles match the blood from **Suspect A's shoes?** ____Yes____

   If so, list them:

   2, has the same DNA

3. Did any of the profiles match the blood from **Suspect B's shoes?** ____Yes____

   If so, list them:

   B Shoes match. B, and B match.

4. Did any of the profiles match the blood from **Suspect C's shoes?** ____NO____

   If so, list them:

5. Did any of the profiles match the **blood from the deceased male?** ____Yes____

   If so, list them:

   Blood from Suspect C Shoes.

6. Did any of the profiles match the **blood found on the ground** outside the truck?

   Yes

   If so, list them:

   Suspect C Shoes match the blood.

7. Do these **results** make any individual a more likely suspect than the others?

   Yes

   Explain your answer, citing facts from the DNA analysis.

   When you have matching DNA that matches the Decessed, that Shoes you have been around them.

8. What other types of evidence would be helpful to be sure you've found the murderer?

   Finger prints, time they have been around the Suspect.

Check your answers with your instructor before you continue.

ACTIVITY 3   SHORT TANDEM REPEAT (STR) ANALYSIS

Frequently, the amount of DNA collected at crime scenes is insufficient for RFLP analysis. For this reason, forensic scientists began to use a procedure called the **polymerase chain reaction (PCR)**. PCR allows the scientist to **target specific chromosome locations and make copies of the DNA.** By using this technique, scientists can copy samples containing a small amount of DNA (like the back of a licked envelope).

Scientists currently use noncoding **DNA loci** called **short tandem repeats (STRs)** in forensic DNA testing. **An STR is a small sequence of bases that is repeated.** The number of times that the unit is repeated varies from person to person. For example, an STR with the **base sequence** AATC may be repeated between 12 and 24 times.

As with all of your DNA, the alleles for this example STR locus would be inherited from your parents.

Each individual would have **two alleles for this STR locus** (recall that each person inherits one allele from each parent).

So, an individual's **genotype for this locus could be 13, 14 (meaning that the allele this person inherited from one parent repeats the STR base sequence 13 times and the other allele repeats 14 times).**

To make an **STR profile,** the DNA sample is tagged with a **fluorescent label.** The label is detected by a **laser,** which sends a **signal to a computer.** This signal produces a peak on a computer printout.

The pattern of peaks on an STR profile (see **Figure 20-7**) represents the alleles at each locus (recall that there are two alleles at each locus). If you see only **one peak** for a locus, the individual has **two alleles that are alike. They are homozygous** at that locus.

In the sample locus to the right, note that there are **numbers below each peak.**

These two numbers are used to designate the person's **genotype.**

In this example, the genotype is **13, 14.**

The 13 allele has the STR "phrase" AATC repeated 13 times. The 14 allele has the STR phrase AATC repeated 14 times.

The **letter and number above the peaks** are **abbreviations** that represent the different STR loci.

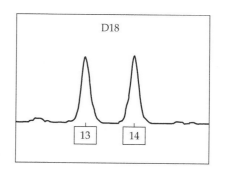

FIGURE 20-7.   Computer Printout of Sample Alleles at One STR Locus

As with RFLP, an STR profile doesn't consist of only one locus. **Multiple loci are used to develop a DNA profile.** The forensic community has adopted **13 core STR loci** that are tested by all forensic laboratories. By having all labs perform the same test, results from around the country can be compared. These results are currently being stored and compared in a **computer database system** called **CODIS (Combined DNA Index System).**

DNA samples are frequently analyzed in **two STR sets. Group 1** contains **nine STR loci** plus one locus used for gender identification, abbreviated **AML (XX or XY).**

Group 2 contains **the remaining four STR loci, plus the AML locus, plus two STR loci from Group 1 that are repeated** as a **control in Group 2,** to ensure that samples originated from the same individual.

A complete STR profile is shown in **Figure 20-8.**

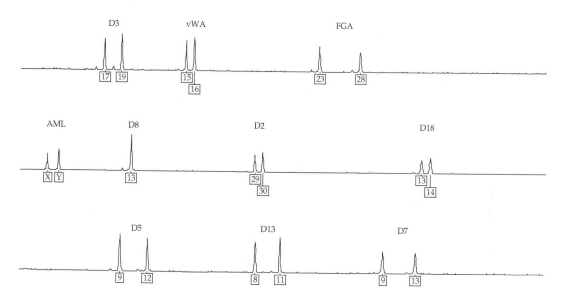

Group 1

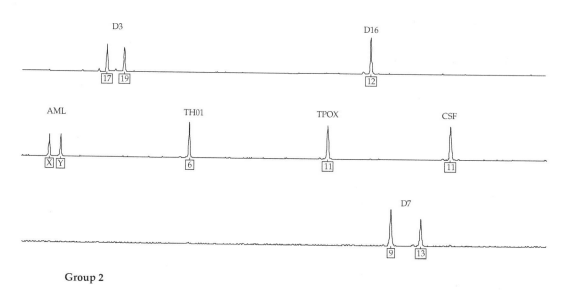

Group 2

FIGURE 20-8.   Example of an STR Profile

✓ Comprehension Check

1. **(Circle one answer.)** This DNA profile was taken from a **male / female. Explain** your answer.

2. Using the STR abbreviations on the profile, which two loci are the **controls** (repeated in both Groups 1 and 2)?

3. How many alleles in **Group 1** of the STR profile are homozygous? _____

    How many are heterozygous? _____

    How many alleles in **Group 2** of the STR profile are homozygous? _____

    How many are heterozygous? _____

    Explain your answers.

4. What do the numbers 9 and 13 designate in the **D7 locus** of Group 2?

5. **Challenge Question!** From your answer to Question 3, can you tell if any of these alleles are dominant or recessive? **Explain** your answer.

Check your answers with your instructor before you continue.

ACTIVITY 4

SOLVING A CRIMINAL CASE USING STR EVIDENCE

The local police department responded to a call regarding a sexual assault. Upon arriving at the scene, they found the victim lying on the floor, partially clothed, and crying. The victim, Jane Doe, was transported to the hospital for treatment and evidence collection. Using a brief description of the suspect, the police apprehended two individuals, but the victim wasn't able to make a positive identification.

At the hospital, **two evidence samples** were collected from the body of the victim. The evidence samples **tested positive for semen** and were sent to the lab for STR analysis. In addition, **known blood samples from all individuals** were obtained and submitted for STR analysis.

1. Interpret the STR profiles in **Figures 20-9** through **20-13** on the following pages. Write the genotypes in the STR table on the following page.

Group 1 Sample	D3	v WA	FGA	AML	D8	D2	D18	D5	D13	D7
Semen sample one										
Semen sample two										
Victim										
Suspect one										
Suspect two										

Group 2 Sample	D3	D16	AML	TH01	TPOX	CSF	D7
Semen sample one							
Semen sample two							
Victim							
Suspect one							
Suspect two							

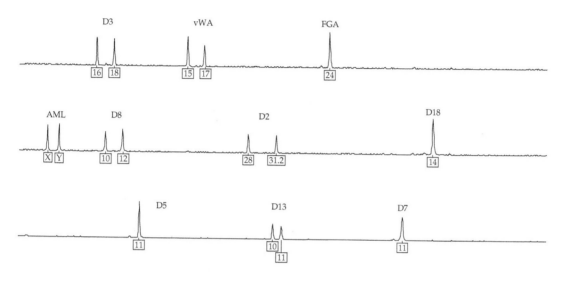

Group 1

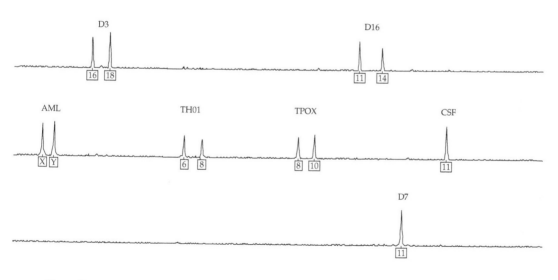

Group 2

FIGURE 20-9.   STR Profile of Semen Sample One

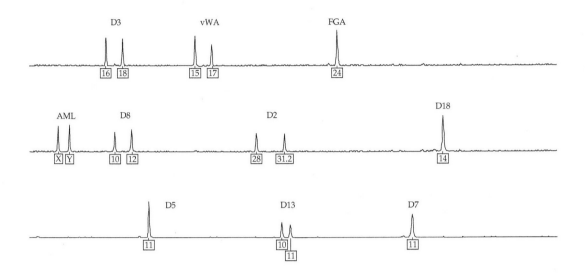

Group 1

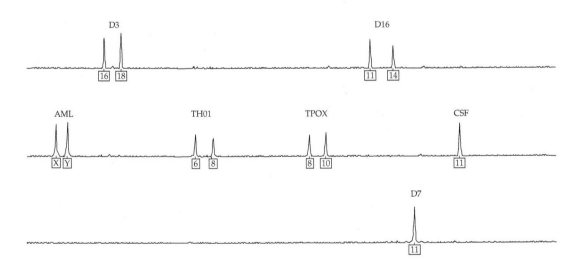

Group 2

FIGURE 20-10.   STR Profile of Semen Sample Two

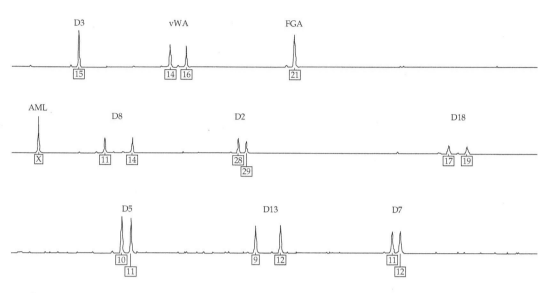

Group 1

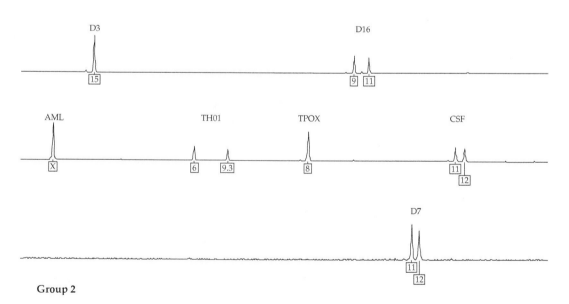

Group 2

FIGURE 20-11. STR Profile of Known Blood Sample from the Victim

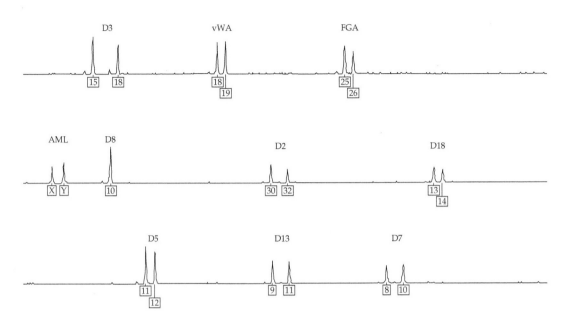

Group 1

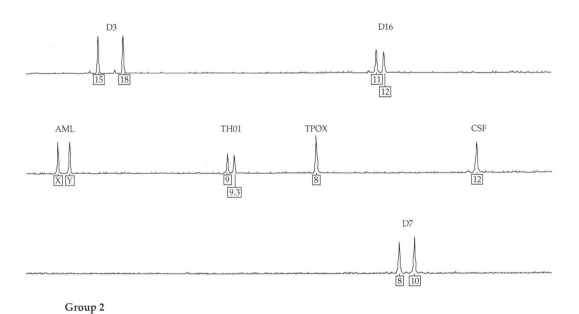

Group 2

FIGURE 20-12. STR Profile of Known Blood Sample from Suspect One

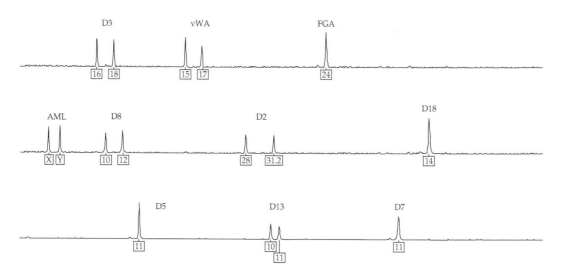

Group 1

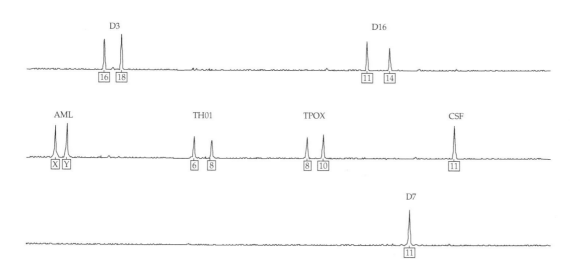

Group 2

FIGURE 20-13.   STR Profile of Known Blood Sample from Suspect Two

> ### Note:
>
> **When trying to determine whether two STR profiles are a match, remember that the alleles at 100% of the loci have to be the same. If even one locus doesn't match, the profiles are not from the same person.**

✓ Comprehension Check

1. Based on the results of your analysis, what conclusions can you draw about this case? **Be specific. Support your answer with facts** from the STR profiles.

2. The detective investigating this case asks you the following question. Please respond.

   "If the DNA profile from the evidence matches the suspect, he must be guilty of sexual assault, right?"

3. **(Choose one answer.)** When a blood sample is submitted for STR analysis, the DNA tested is taken from **red blood cells / white blood cells / platelets. Explain** your answer.

4. What results would you expect from a comparison of STR profiles taken from the following? **Explain** your answers.

    a.   Identical twins:

    b.   Fraternal twins:

Check your answers with your instructor before you continue.

ACTIVITY 5

DECIDING A PATERNITY SUIT INVOLVING STR ANALYSIS

STR profiles can also be used to help identify a child's biological parents. As you recall from the lessons on meiosis, a child inherits half of its genetic information from each parent. This makes matching of DNA profiles more challenging.

If a child has alleles in his/her STR profile that don't match those of the mother, these alleles must have been inherited from the father.

A young woman claims that her two-year-old son is the child of one of the members of the hit rock group "First Impressions." During the time of conception, she was dating all three of the band members.

In her sworn testimony, the mother states, "The father is one of the band members. They were the people I spent time with during that whole year. The problem is, I don't know which musician is the biological father of my baby."

None of the members of the band have admitted to fathering the child. During the investigation, all members of the rock group agreed to provide DNA samples for analysis. You're the forensic scientist appointed by the court to analyze the DNA evidence.

The profiles for all parties involved are summarized in the following STR table.

1. Compare the child's STR profile with the mother's. In the child's profile, **cross out** any **Group 1 and Group 2** alleles that **are the same as the mother's** (as shown in the following example).

2. Next, compare the alleles of each of the potential fathers with the child's profile. **Circle all matching alleles.**

3. Based on your analysis of the STR profiles, which of the men could be the father of the child? **Explain** your answer, using facts from the STR analysis.

> mike, miller, has the only one with two
> cirles that maten the Son and mother.

Check your answers with your instructor before you continue.

Group 1	D3	vWA	FGA	AML	D8	D2	D18	D5	D13	D7
Child	13, 18	12, 12	20, 25	X, X	9, 19	28, 30	10, 12	7, 16	13, 13	6, 9
Jenny Smith	12, 18	12, 15	20, 20	X, X	9, 19	28, 32	12, 19	12, 16	13, 14	6, 6
Peter Niles	13, 13	11, 12	22, 25	X, Y	9, 9	31, 32	9, 10	10, 12	8, 9	9, 11
Phillip Cruze	12, 13	12, 20	25, 29	X, Y	10, 19	30, 30	10, 19	7, 9	11, 13	9, 15
Mike Miller	13, 19	12, 12	24, 26	X, Y	9, 15	31, 31	14, 15	13, 16	15, 15	6, 8

Group 2	D3	D16	AML	TH01	TPOX	CSF	D7
Child	13, 18	11, 14	X, X	9, 11	7, 11	13, 14	6, 9
Jenny Smith	12, 18	8, 11	X, X	10, 11	7, 13	6, 14	6, 6
Peter Niles	13, 13	9, 13	X, Y	5, 7	8, 9	12, 14	9, 11
Phillip Cruze	12, 13	14, 16	X, Y	9, 10	11, 13	11, 13	9, 15
Mike Miller	13, 19	11, 16	X, Y	7, 11	8, 12	9, 10	6, 8

Self Test

1. What is meant by "noncoding" DNA regions?

2. How can noncoding DNA regions be used to determine a person's identity? **Be specific.**

3. How is comparing STR profiles to determine paternity different from trying to match a suspect to a blood sample?

4. You're a park ranger at Yellowstone National Park. You've discovered the internal organs of a deer just inside the park boundaries. Because hunting in the park is illegal, you notify the local police to be on the alert for a recently killed deer. The police spot a truck carrying the carcass of a large deer a few miles outside the park. They suspect this is the deer that was killed inside the park boundaries, but they need some evidence linking this deer to the remains found in the park.

    What test(s) could determine whether the deer in the truck was killed inside Yellowstone Park? **Explain** your answer.

5. On Saturday, September 28, a young boy was bitten by a large black dog on the corner of Fifth Avenue. Investigating the situation further, police officers learned that a neighbor has a large black dog that often barks ferociously at people walking down the street. The owner maintained that the dog made a lot of noise, but was basically friendly. The officers, however, discovered a substance, which appeared to be blood, on the dog's collar. The owner states that he cut his hand on the dog collar and that the blood is his.

The boy's family decides to sue the dog's owner. They wish to be compensated financially and also have the dog destroyed. However, the dog's owner is adamant that his dog never left his fenced yard on the date in question. You're a forensic scientist for a private firm that has been asked to evaluate the case for the civil trial. All samples have been tested and found to be human blood. Interpret the results of the DNA testing as presented in **Figures 20-14** through **20-16**.

Summarize your findings in this investigation. **Support your statements with evidence** from the DNA fingerprints.

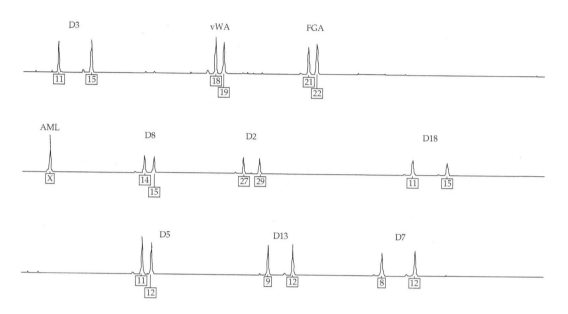

Group 1

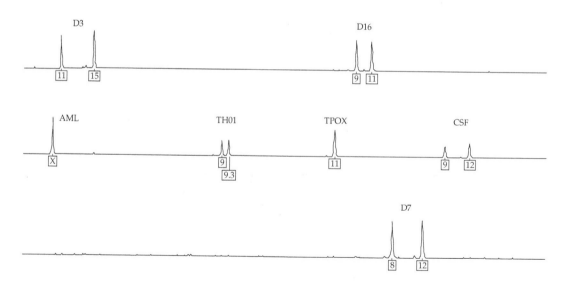

Group 2

FIGURE 20-14. Dog Owner's Blood

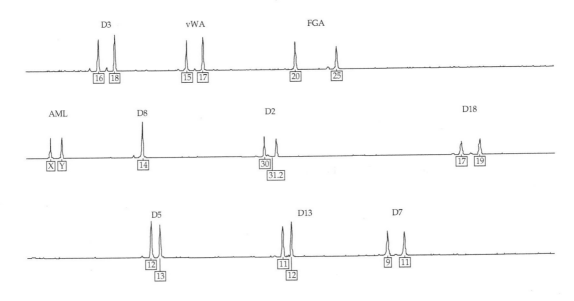

Group 1

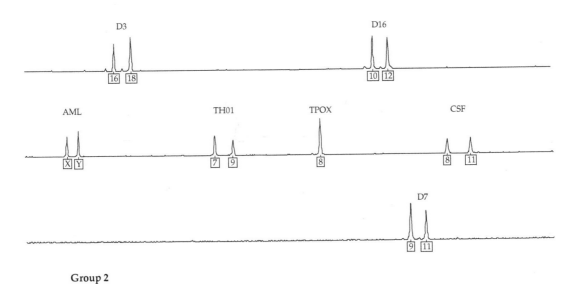

Group 2

FIGURE 20-15. Boy's Blood

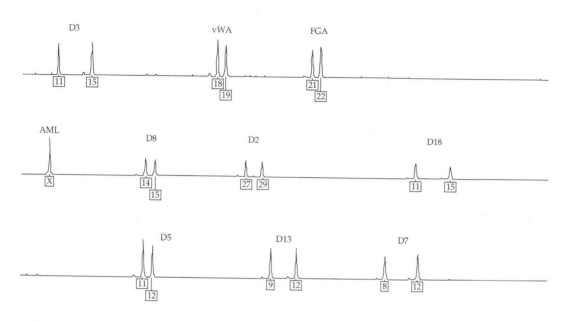

Group 1

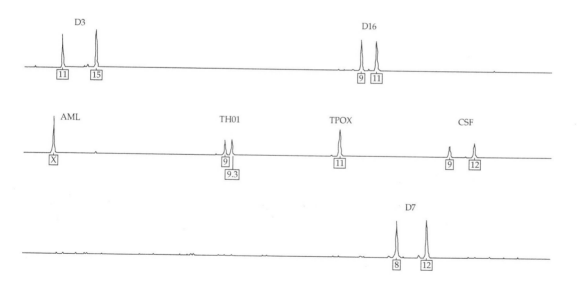

Group 2

FIGURE 20-16.   Blood from Dog's Collar

Using Biotechnology to Assess Ecosystem Damage

Objectives

After completing this exercise, you should be able to:

- explain the effects of environmental factors on bioluminescence and give examples
- correctly prepare and use serial dilutions
- demonstrate awareness of factors that can affect the accuracy of experimental results
- explain the utility of serial dilutions to analyze the effects of commonly used chemicals on ecosystems
- use experimental results to rank products in terms of their potential to cause harm to human health or the environment
- draw graphs that present data clearly and accurately
- interpret data in tables, charts, and graphs
- apply your knowledge of chemical toxicity to real-life situations

CONTENT FOCUS

Many types of harmful compounds are released into the environment by industry, by agriculture, and even from our own households. It's important not only to detect harmful materials but also to predict their effects on the environment. This is difficult because pollutants are often released in **small amounts, which rapidly disperse** into the soil or water. It's also important to identify **which types** of chemicals are present in the environment, and whether these chemicals are **harmful** to living organisms.

To address this problem, Dr. Kenneth Thomulka, a scientist from Philadelphia, developed an inexpensive, convenient field-testing method to discover whether toxic chemicals are present in the environment.

The method uses harmless marine bacteria as **biological indicators** for the presence of toxic chemicals.

ACTIVITY 1   EXPLORING BIOLUMINESCENCE

The ability of an organism to produce light is called **bioluminescence.** Bioluminescence is exhibited in a variety of organisms, from bacteria to fireflies in your backyard. It is the only light source for marine organisms living deep in the ocean. Like fireflies, some deep-sea fishes use their lights as signals to find mates. Others, like the deep-sea anglerfish, wave glowing lures to attract smaller fish as prey.

Many bioluminescent organisms, including bacteria, have the enzyme **luciferase.** Luciferase and **oxygen** are needed for the complex chemical reactions used to produce light.

Note:
Read these instructions COMPLETELY BEFORE the lights go out!

1. Work in groups. Get the following supplies: **a test tube rack and one 15-ml screw-capped test tube containing a sample of the bacterial culture.**

2. Place the test tube in the rack and set it on your laboratory table. **Unscrew and remove the lid.** Leave the test tube **undisturbed** for 15 minutes.

Caution!
At times during this laboratory period, the room will be completely dark. Make sure the aisles and working areas around your table are clear.

3. To demonstrate the production of light, the class will observe a flask containing **bioluminescent marine bacteria.** With the **lights out,** observe what happens when the instructor swirls the flask containing the bacteria. **Record** your observations.

4. **Continue to observe** the flask for several minutes as it sits undisturbed on the laboratory counter. **Record** your observations.

5. **Without moving or disturbing the test tube rack,** check the tube of bacteria on your laboratory table. Observe the tube **very carefully.**

   Where is the light level **most** intense? _____

   Where is the light level **least** intense? _____

   Using the information presented in the **introduction** to this exercise, which mentions **two** substances needed for bioluminescence to occur, **explain** your observations.

6. **Rank the level of light produced** by the bacteria on a scale of **0 (no light) to 4 (most light).**

    Top of the tube _____

    Middle of the tube _____

    Bottom of the tube _____

7. **Shake** the test tube **gently** and observe the results. **Describe** what you see.

8. After shaking the test tube, **rank the level of light produced** by the bacteria on a scale of **0 (no light) to 4 (most light).** _____

    Why did the change in light level occur?

Bacterial test kits are commercially available for detecting pollutants in aquatic ecosystems, such as lakes, streams, and oceans. Manufacturers of household cleaners, shampoos, and cosmetics have started using bacterial testing to replace the more traditional, but highly controversial, tests done with rabbits, rats, and mice.

In the following experiments, you'll use bioluminescent bacteria to evaluate the toxicity of common household products. You'll try to determine whether any of these household products can be damaging to the environment if they **aren't used and disposed of properly.**

Any environmental condition that's harmful to the bacteria's health will cause a decrease in light production.

The point of this experiment ISN'T to see how well you can kill bacteria. These bacteria are not harmful. They are valuable and necessary for the environment. The point is to find out how you can PREVENT household products from harming organisms in the environment, including bacteria.

ACTIVITY 2   GETTING STARTED

1. Work in groups. Get the following supplies: **a small flashlight with a red filter, one small beaker, one graduated 10-ml pipette with manual dispenser, and one graduated 1-ml pipette with manual dispenser.**

    You'll also need the following: **six large test tubes in a rack, six small test tubes in a rack, scissors, masking tape, a dispenser bottle filled with 3% saline solution, a tray for used glassware, and a container for waste fluids.**

2. Fill the small beaker half full with **tap water.** Attach the manual dispenser to the end of a **graduated 10-ml pipette.**

    Practice filling and dispensing fluid from the pipette following the directions given by your instructor.

3. Now that you're an expert pipette user, you are ready to set up your experiment.

    With masking tape, label the **large** test tubes **D1** through **D6.**

    Make sure you place the tape labels as **close as possible to the top of each tube.**

4. The following household products are available for testing:

drain cleaner	automobile antifreeze
mouthwash	daily shower cleaner
household cleaner	herbicide (weed killer)
toilet-bowl cleaner	automobile wheel cleaner

    ### Caution!
    Some of these products may be harmful to your skin or eyes. Be careful not to spill any on your hands as you measure. If a spill occurs, DON'T touch your face or eyes. **Wash your hands thoroughly before proceeding.**
    ---

5. Your instructor will assign a household product to your group.

    Obtain a small container filled with your assigned household product.

    Enter the name of the product your group will be using: _____

ACTIVITY 3   MAKING ACCURATE SERIAL DILUTIONS

1. You'll **dilute** the chemical with various amounts of saline to test the reaction of the bacteria to **different strengths** of this product.

    The dilutions will produce a series of proportionally weaker solutions, commonly referred to as **serial dilutions**. In this manner, you can determine what dilution (concentration) of this chemical is harmful to the bacteria.

    In this experiment, we'll compare products to see which are more harmful than others and we'll base our assessment of toxicity on the level of dilution necessary to dispose of the chemical safely.

    Rather than just diluting randomly to make this assessment, it's more useful to dilute the chemical in **measured steps (making a serial dilution).** By exposing bacteria to different concentrations (dilution levels) of the various household products, you can determine the exact level of dilution that will make the product harmless.

    Since these are marine bacteria, you'll make your dilutions with **3% saline** (a mild salt solution), which simulates the water in the bacteria's natural environment.

2. The dilution levels you'll use are listed in **Table 21-1.**

TABLE 21-1 DILUTION SEQUENCE		
TEST TUBE NUMBER	DILUTION	PERCENT CONCENTRATION OF PRODUCT
D1	Full strength	100
D2	1:10	10
D3	1:100	1
D4	1:1,000	0.1
D5	1:10,000	0.01
D6	No product added	0

Hint:
To obtain valid results, you must be very careful to make accurate measurements.

3. Using the **10-ml graduated pipette,** transfer **9 ml** of your assigned household product into the first **large** test tube (marked **D1**). This will be the **full-strength** tube.

> ## Note:
> **Place all used pipettes and glassware into the container provided.**

---

4. Using a **1-ml graduated pipette,** transfer **1 ml** of household product from the container with the original sample into the next large test tube (marked **D2**).

5. To the **same** tube **(D2),** add **9 ml of saline** solution from the stock bottle. **To dispense the saline,**

    ■ hold the test tube under the spout

    ■ **slowly** raise the pump handle as far as it will go

    ■ **gently** push the handle down to dispense

    Without spilling the contents, **shake the test tube gently** to mix the solution. This will be the **1:10 dilution.**

6. **From test tube D2, remove 1 ml** of solution and transfer it to test tube **D3.**

    Following the instructions above, add **9 ml of saline** to tube **D3.**

    Shake gently to mix the contents. This will be the **1:100 dilution.**

7. **From test tube D3, remove 1 ml** of solution and transfer it to test tube **D4.**

    Following the instructions above, add **9 ml of saline** to tube **D4.**

    Shake gently to mix the contents. This will be the **1:1,000 dilution.**

8. **From test tube D4, remove 1 ml** of solution and transfer it to test tube **D5.**

    Following the instructions above, add **9 ml of saline** to tube **D5.**

    Shake gently to mix the contents. This will be the **1:10,000 dilution.**

    Remove 1 ml of solution **from test tube D5** and dispose of it in the waste container.

9. To test tube **D6,** add **only 9 ml of saline.** No household product will be added to this tube.

✓ Comprehension Check

1. Which test tube contains the **weakest** sample of household product? _____

2. Which test tube contains the **strongest** sample? _____

3. Which test tube is the **control?** _____

4. Why is a control tube needed in this experiment?

5. Why was 1 ml of liquid **removed from test tube D5?**

6. **In your own words,** explain the **purpose of making a serial dilution** for this experiment.

Check your answers with your instructor before you continue.

ACTIVITY 4   PREPARING BACTERIAL SAMPLES

1. Get the following supplies: one **clean 1-ml pipette.**

   Using the tube of bacterial culture you looked at during Activity 1, transfer **1 ml of bacteria** to each of the **six small test tubes.**

2. To **each** of the six small test tubes, add **9 ml of saline** solution and set them aside.

> ### *Wait!*
> **All laboratory groups must start the next part of the experiment together! While you're waiting, read the instructions for Activity 5 COMPLETELY.**

ACTIVITY 5   TESTING THE TOXICITY OF YOUR CHEMICAL

1. **Record the time** when you begin this procedure. _____

2. **When instructed, carefully empty one small test tube of bacteria into each of the large dilution tubes.**

   Without spilling, **shake each tube gently** to mix the contents.

   Place the empty "bacteria" tubes into the **waste container.**

   It is important to remove these empty tubes from your workspace because the light pollution from the bacteria remaining in the empty tubes will interfere with your observations of the light levels in your experimental tubes.

3. The bacteria/chemical combination must incubate for **15 minutes.**

   Record the time when the incubation period will be completed. _____

4. When the incubation period is completed, observe the bacteria and evaluate the level of bioluminescence present.

Caution!
The room will be COMPLETELY DARK for this activity. Make sure your test tube rack, paper, pen, and flashlight are conveniently positioned before the lights go out!
It will take several minutes for your eyes to become adapted to the dark. To avoid accidents during this period, don't move around.
The flashlight will be only turned on briefly to record your results!

5. **Rank the light** produced by your bacteria according to the **0 through 4 scale** you used in Activity 1 and record your results in **Table 21-2**.

TABLE 21-2		
LIGHT INTENSITY OF BACTERIA EXPOSED TO SERIAL DILUTIONS OF ONE HOUSEHOLD PRODUCT (0–4 SCALE)		
TEST TUBE NUMBER	DILUTION	RANKING (0–4)
D1	Full strength	
D2	1:10	
D3	1:100	
D4	1:1,000	
D5	1:10,000	
D6	No product added	

6. **Record** the results for your group on the **master chart** at the front of the room.

   From the master chart, copy the results for the entire class into **Table 21-3.**

TABLE 21-3

LIGHT INTENSITY OF BACTERIA EXPOSED TO SERIAL DILUTIONS OF ALL EIGHT HOUSEHOLD PRODUCTS (0–4 SCALE)

TUBE #	DRAIN CLEANER	MOUTHWASH	HOUSE-HOLD CLEANER	ANTIFREEZE	SHOWER CLEANER	WEED KILLER	TOILET-BOWL CLEANER	AUTO WHEEL CLEANER
D1 (full strength)								
D2 (1:10)								
D3 (1:100)								
D4 (1:1,000)								
D5 (1:10,000)								
D6 (control)								

7. Using the data from **Table 21-3,** make **eight bar graphs** that show the results of each experiment in **Figure 21-1.**

   Label each graph with the name of the household product that was tested.

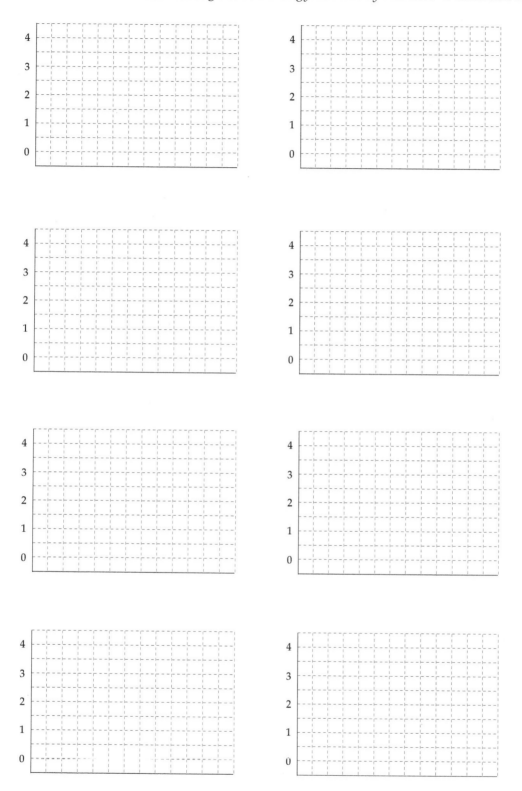

FIGURE 21-1.   Comparison of Test Results for Eight Household Products

Check your graphs with your instructor before you continue.

✓ Comprehension Check

1. Which household product was the **most toxic?** _____

   Why did you conclude that this product was the most toxic? In your explanation, **include facts** collected during the bacteria experiments.

2. What part(s) of the food chain might be affected if this product was disposed of improperly in the environment? **Explain** your answer.

3. Imagine this situation. A type of drain cleaner is found to be harmless to bacteria if it is diluted in a 1:10,000 ratio. A 1:10,000 dilution is the same as **1 ml of chemical diluted by 10 liters of water.** The drain cleaner container holds 200 ml of chemical. If the directions tell you to pour the entire contents of the container down the drain, how **many liters of water** would be needed to dilute this chemical to safe levels?_____ liters   **Show your work.**

4. In an experiment testing the toxicity of a car wash product, bacteria exposed to the product at a **1:1,000 dilution** showed a light level of **two.** Bacteria exposed to a **1:10,000 dilution** of the same product showed a light level of **three.**

    In order to dispose of this product safely, is a 1:10,000 dilution sufficient? **Explain** your answer.

5. What would you conclude if **two of the tested products** showed **NO bioluminescence** at a **1:1,000 dilution? Explain** your answer.

6. What would you conclude from your experiment if **two** of the tested household products showed the **same light intensity at a 1:10,000 dilution? Explain** your answer.

7. **Challenge Question!** Can the bioluminescence of bacteria exposed to a **1:10,000 dilution** of a chemical ever equal the amount of bioluminescence in the **control tube? Explain** your answer.

Check your answers with your instructor before you continue.

Self Test

Figure 21-2 contains the results of a bioluminescent bacteria assay performed with two lawn-care products that frequently find their way into sewers, streams, lakes, and ponds.

1.   At which dilution level was **Product A** most toxic? _____

     What is the **lowest** dilution level at which **Product A** was harmless? _____

2.   At which dilution level on the graph is **Product B** harmless? _____ **Explain** your answer.

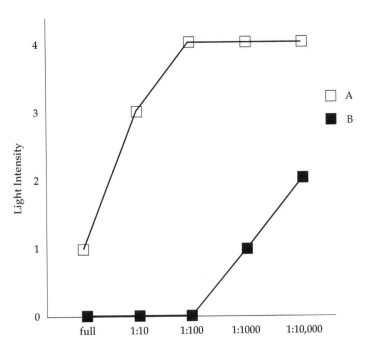

FIGURE 21-2.   Bioluminescent Bacterial Assay of Two Lawn-Care Products

3. If these two lawn-care products do the same job, which product is less harmful to the environment? **Explain** your answer.

Self Test Answers

Exercise 1 — Introduction to the Scientific Method

1. B — state hypothesis

2. C — state results (facts only)

3. A — test hypothesis (by experiment or observation)

4. A — test hypothesis (by experiment or observation)

5. B — state hypothesis

6. B — state hypothesis

7. C — state results (facts only)

8. a. The points are too closely spaced together (they should be spread out as much as possible along the Y-axis) and crowded into the corner (they should be evenly spaced along the X-axis).

   b. No numbers or information about scale on the X-axis

   c. No graph title

9. a. Titles missing for the X- and Y-axes

   b. Numbers on the Y-axis do not have equal intervals; the scale changes from intervals of 50 (50–100) to intervals of 500 (100–500)

   c. No numbers or information about the scale on the X-axis

   d. Plotted points extend above the number scale on the Y-axis

10.   a.   Titles missing for the X- or Y-axes

      b.   No numbers or information about the scale on the X-axis

11.   Decreased

12.   About 13%

13.   About 88%

14.   1986

Exercise 2 — Using the Compound and Dissecting microscopes

1.

Ocular lens (eyepiece)	Allows you to view the specimen; magnifies image 10 times (10×); contains pointer
Stage	Holds the specimen; has a mechanism to move the slide around
Condenser lens	Focuses light on the specimen
Nosepiece	Revolves to change objective
Scanning lens (4×)	Objective lens used to first locate a specimen
Iris diaphragm	Regulates the amount of light that passes through the specimen
Fine focus knob	Moves objective lenses in small increments for small adjustments in image clarity
Scanning lens (4×)	Objective lens with the lowest magnifying power
High power (40×)	Objective lens with the highest magnifying power
Coarse focus knob	Moves objective lenses rapidly in large increments for initial adjustment in image clarity

2.   Ocular lens = 15   Objective lens = 20

         Total magnification: $15 \times 20 = 300$

3.

   ■   In the compound microscope, the light source is beneath the specimen. Therefore, only thin specimens can be viewed.

   ■   The dissecting microscope can provide light from many different directions so that large, thick objects can be viewed.

   ■   In the compound microscope, the image is viewed upside down and backward. This is not true for a dissecting microscope.

   ■   The dissecting microscope has a zoom lens that can gradually increase magnification. The lenses of the compound microscope have fixed magnification levels.

   ■   The compound microscope has much greater magnification ability than a dissecting microscope.

4.   Compound — cells from the lining of your stomach

         Dissecting — a splinter in your finger

         Dissecting — a cockroach you found in the kitchen

         Compound — mold from your shower curtain

5.   The daphnia moved away from you and to your left (because images in the compound microscope are inverted and reversed).

6.   a.   Go back to the 10× lens and center the specimen in the middle of the field of view or check to make sure the 40× objective is clicked into position.

     b.   Adjust the iris diaphragm to let in more light.

     c.   Check to make sure the objective lens is clicked into position, or check to see if the objective and ocular lenses are clean.

     d.   Use lens paper to clean the slide, the objective lenses, and the ocular lens.

     e.   These are air bubbles. Lift the cover slip and lower it gently at a 45° angle to expel the air.

Exercise 3 — Functions and Properties of Cells

1.   Cell wall, chloroplast, central (sap) vacuole

2.   Arrow should show water moving from the bag into the beaker.

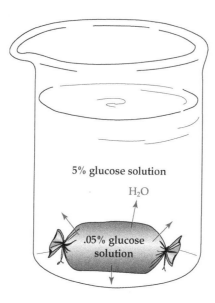

5% glucose solution

H_2O

.05% glucose solution

3.   **Dialysis** is the separation of different sized molecules through a selectively permeable membrane. As diffusion takes place across the membrane, the composition of the blood changes. Waste products such as phosphate ions, urea, and potassium ions cross the membrane into the dialysis fluid. Blood cells, proteins, and other large molecules cannot cross the membrane and are retained in the blood.

4.   Distilled water has no dissolved solutes. Osmosis refers to the movement of water molecules from an area of high concentration to an area of lower concentration. In the described situation, there is a **lower concentration** of water molecules inside the cell (in the cytoplasm) and a **higher concentration** of water outside the cell (the distilled water). Through the process of **osmosis,** the distilled water will cross the **cell membrane** and enter the blood cells, causing them to swell and rupture the cell membrane.

5.   The sugar cube will dissolve faster in the hot tea. During **diffusion,** molecules move from an area of high concentration to an area of lower concentration. Since the water is hot, molecular movement increases. **Collisions** between molecules become more frequent. As molecules bump into each other, they **diffuse** outward, spreading from the area of **high sugar concentration** (the sugar cube) to an area of **lower sugar concentration** (the cup).

6.   The smoke **molecules** are moving by **diffusion** over the restaurant barrier from an area of **high smoke concentration** (the smoking section) to an area of **lower smoke concentration** (the nonsmoking area).

7. Excess salt in meltwater can cause osmotic stress in plants growing nearby. As the salt water seeps into the soil, it surrounds the root cells with a hypertonic solution, causing dehydration.

Exercise 4 — Investigating Cellular Respiration

1. Bromothymol blue turns yellow when CO_2 is present. Since the solution did not change color, we can conclude that cellular respiration did not take place in the container (no CO_2 was produced). In this situation, we cannot even be sure the green spots were plants.

2. a. The cloudy result indicates that the gas obtained from the body cavity contained CO_2.

   b. CO_2 is a by-product of cellular respiration. Cellular respiration occurs only in living organisms. Since the animal is dead, the CO_2 produced must be coming from the cellular respiration of other organisms (decomposers present on the carcass).

3. A — aerobic respiration

   Aerobic respiration produces about 36 ATPs per glucose molecule, whereas anaerobic forms of cellular respiration produce only 2 ATPs per glucose molecule.

4. Bromothymol blue returned from yellow to its original blue color because CO_2 was no longer present. Since plants absorb CO_2 during photosynthesis, and the plant was exposed to sunlight during the several hours in question, this is evidence that photosynthesis occurred.

5. B — only under anaerobic conditions

   In the absence of sufficient oxygen, yeast will shift their method of cellular respiration to an anaerobic method — alcohol fermentation. In this way, they can continue to obtain ATP to support cellular activities.

6. C — under both aerobic and anaerobic conditions

   Breakdown of glucose for cellular respiration begins with glycolysis, an anaerobic process. Cellular respiration will continue as shown in **Figure 4-3,** but the pathway used is determined by the presence or absence of oxygen.

7. B — only under anaerobic conditions

   During strenuous exercise, when sufficient oxygen is not available to muscle cells, cellular respiration shifts to an anaerobic method — lactic acid fermentation. In this way, muscle cells can continue to obtain ATP to support cellular activities.

8.   A — aerobic conditions only

       Under ideal conditions, cells can maximize their ATP production.

9.   D — Even though anaerobic respiration produces small amounts of ATP, the amount is not sufficient to meet the energy needs of an animal for long periods of time.

Exercise 5 – Enzyme Activity

1.   We have seen that unusually high temperatures can affect the three-dimensional folding of a protein. In enzymes, if the structure of the active site is altered, the ability of the enzymes to function may be destroyed. High fevers may affect enzymes throughout the body.

2.   Enzymes are sensitive to changes in environmental conditions and many are quite specific as to he pH range in which they can function.

3.   3.5

4.   At pH 3.7, it took 14 minutes before noticeable enzyme activity was observed. In comparison, enzyme activity was observed after only 2 minutes at pH 3.5.

5.   The enzyme is specific because its optimal function was centered around pH 3.5. Enzyme activity at pH values even slightly above or below 3.5 was greatly decreased.

6.   The production of melanin (and, thus, the fur color in Siamese cats and Himalayan rabbits) is related to body temperature. This trait does not apply to the colors of skin, fur, and feathers in all animals, just these specific breeds.

7.   Blood circulation and body temperature both decrease in the extremities. Since body temperature is lower in the paws, ear tips, and nose, the enzyme for melanin production is active. Core temperatures at the center of the body are too high for melanin synthesis to be completed.

Exercise 6 — Food Analysis and Choices for Good Health

1.   Your morning orange juice doesn't contain protein.

2.   Since potatoes contain large amounts of starch, the iodine would probably turn black (a positive starch test).

       Since animal tissues do not contain starch, the iodine test should be negative (no color change).

3.   Peanut butter   A seed; contains the embryo and food reserve

      Tuna   Skeletal muscle tissue from a fish; produces voluntary
                             movements

      Refried beans   A seed; contains the embryo and food reserve

      Milk   Liquid high in nutrients; produced by mammary
                             glands of mammals to feed their offspring

      Hamburger   Skeletal muscle tissue from a cow; produces voluntary
                             movements

      Lettuce   Leaf of lettuce plant; site of photosynthesis

4.   70 mg Sodium

5.   500 kcal

6.   4.5 grams

7.   24 kcal

8.   4 servings

Exercise 7 — The Skin: Example of an Organ

1.   I — sebaceous gland

2.   J — dermis

3.   A — keratin

4.   D — epithelial cells

5.   C — erector muscle

6.   H — subcutaneous layer

7.   B — melanin

8.   E — epidermis

9.   No. Lotions applied to the outer skin surface do not penetrate to the lower layers
      of the epidermis where new skin cells are produced.

10. The dermis. The nicotine medication must penetrate through the epidermis and enter blood vessels in the dermis.

11. On the sole of the foot, the epidermal layer is thickened and calluses may be present. Penetration of the patch medication through to the dermis might prove difficult.

12. In second-degree burns, the epidermis has been destroyed, exposing the sensitive nerve endings in the dermis. Pain will be severe. In third-degree burns, nerve endings are completely destroyed. Consequently, the burn patient feels no pain until the nerves begin to regenerate and repair themselves.

Exercise 8 — The Musculoskeletal System

1. D — yellow marrow

2. E — red marrow

3. G — compact bone

4. F — smooth muscle

5. C — cardiac muscle

6. B — skeletal muscle

7. B — skeletal muscle

8. A — spongy bone

9. H — shaft

10. G — compact bone

11. Your friend is experiencing muscle fatigue. The actively contracting muscles become weaker as time passes. This can be caused by lack of ATP, insufficient oxygen, depletion of energy reserves in the muscle cells, and accumulation of metabolic wastes.

12. In the intense heat, the protein fibers in the bone matrix were destroyed. Protein fibers make an important contribution to the strength and resiliency of a bone. Without them, the bone becomes brittle and easily crushed.

13. Tendon

    Ligament

Exercise 9 – Examination of Skeletal Structure

1. K – patella

2. D – fibula

3. O – sternum

4. C – femur

   K – patella

   Q – tibia

5. Q – tibia

6. N – scapula

7. A – clavicle

8. O – sternum

9. Q – tibia

10. E – humerus

11. P – tarsals

12. J – phalanges

13. R – ulna

14. L – radius

15. G – hip bones

16. B – carpals

17. I – metatarsals

    J – phalanges

    P – tarsals

18. The cartilage provides a flexible connection between the ribs and the sternum that can help protect against ipact to the rib cage. In addition, it allows the rib cage to accommodate changes in lung volume during breathing.

Exercise 10 – The Nervous System

1. B – cerebellum

2. J – occipital lobe of the cerebrum

3. F – central nervous system

4. G – medulla oblongata

5. I – white matter

6. K – dorsal root

7. A – ventral root

8. C – corpus callosum

9. E – hypothalamus

10. N – frontal lobe of the cerebrum

11. O – pituitary gland

12. L – meninges

13. Alzheimer's disease affects particularly the frontal and temporal lobes of the cerebrum. Symptoms include memory loss, decreased problem solving ability, and loss of motor skills.

14. The brain stem, which controls basic survival functions, is still operating.

Exercise 11 — Introduction to Forensic Biology

1. C — whorl

2. G — divergence

3. F — bifurcation

4. E — loop

5. J — sweat glands

6. H — fingerprint formula

7. I — points of similarity

8. D — tented arch

9.   The skin is covered with sweat glands that produce perspiration, which can accumulate on the ridges forming the fingerprints. These ridges also accumulate body oils from touching oily surfaces. Together, these leave an invisible impression, which is the fingerprint.

10.   Even though their fingerprints might be the same at birth, differing daily activity cause an accumulation of scars and other marks that can be used to tell the fingerprints of the twins apart.

11.   The blood type is A−. Agglutination with anti-A serum shows the presence of type "A" proteins, but lack of agglutination with anti-B and anti-Rh serums shows that these proteins are absent.

12.   The blood type is AB−. Agglutination occurred with both anti-A and anti-B serums, showing the presence of type "A" and type "B" proteins, but lack of agglutination with anti-Rh serums shows that this protein is absent.

13.   The blood type is O+. Agglutination did not occur with either anti-A or anti-B serums, showing the absence of type "A" and type "B" proteins, but agglutination did occur with anti-Rh serum, which shows that the Rh protein is present.

Exercise 12 — The Circulatory System

1.   1 — tissue capillaries in big toe

6 — pulmonary artery

9 — left atrium

8 — pulmonary vein

4 — right atrium

10 — left ventricle

3 — inferior vena cava

11 — aorta

5 — right ventricle

12 — arterioles

2 — venules

7 — lungs

2. Blood would back up into the right atrium.

3. 96

4. O — tissue capillaries entering big toe

   D — pulmonary artery

   O — left atrium

   O — pulmonary vein

   D — right atrium

   O — left ventricle

   D — inferior vena cava

   O — aorta

   D — right ventricle

   D — tissue capillaries leaving big toe

   O — arterioles

   D — venules

5. Owing to rheumatic fever, scarred heart valves are not closing completely. The hissing sound is caused by a small amount of blood leaking through the valve under pressure when the ventricles contract.

6. Skin arteries dilate (enlarge) to increase blood flow to the skin.

7. right ventricle — 40   pulmonary artery — 40

   pulmonary vein — 100   left atrium — 100

8. Blood flow to the muscles increases from about 21% to about 73%.

   Diameter of blood vessels supplying muscle tissue also increases.

9. Blood flow to the abdominal organs decreases from about 25% to about 3%.

   Diameter of blood vessels supplying abdominal organs also decreases.

10. Blood flow to the skin increases from about 9% to about 12%.

    Increased blood flow to the skin is one mechanism by which the body releases excess heat during exercise.

Exercise 13 — Introduction to Anatomy: Dissecting the Fetal Pig

1.   C — head end of the body

2.   A — back

3.   B — tail end of the body

4.   D — belly side

5.   Female pigs have a genital papilla just ventral to the anus. In addition, the urogenital opening of females is located just below the papilla (as opposed to males, in which the urogenital opening is located just posterior to the umbilical cord).

6.   D — The trachea is the only air passageway to the lungs. Although the esophagus is also located in the throat, it is not involved in the breathing process.

7.   Esophagus

8.   Diaphragm

9.   Under heavy exercise, a person with emphysema will not get enough oxygen for the body's needs. The maximum amount of air moved in and out **(vital capacity)** when breathing deeply is significantly reduced.

10.   When insufficient **oxygen** is transported by **red blood cells,** the **mitochondria** of other body cells cannot function at peak efficiency. The amount of **ATP** produced decreases and, therefore, the amount of **energy** available to power cell activities will also decrease.

Exercise 14 — Organ Systems in the Abdominal Cavity

1.   H — mesenteries

2.   C — villi

3.   D — circular muscle

4.   F — peristalsis

5.   E — longitudinal muscle

6.   B — root hairs

7.   Digestive system: stomach, small intestine, large intestine, pancreas, liver, and gall bladder

      Respiratory system: no organs from this system in the abdominal cavity

Urinary system: kidney, bladder, and ureter

Circulatory system: arteries, veins, and capillaries

8. Cardiac sphincter; stomach acids back up and irritate the lining of the esophagus. Since this region of the esophagus is not far from the heart, this discomfort can be mistaken for chest pains.

9. Stomach acids react with the enamel on the teeth and gradually dissolve it over long periods of time.

10. The malfunctioning organ is probably the bladder. The malfunctioning structure is probably the sphincter associated with the bladder and urethra, which prevents urine from being released accidentally.

11. Design B will be more efficient because more interior surface area is in contact with the air.

12. Crossword puzzle answers:

Exercise 15 – The Reproductive System

1.   (pick up table from question 1 of the self test and complete using the highlighted text)

Structure	Function
testes	production of sperm and sex hormones
follicle	structure in which the egg develops
seminiferous tubules	structure within the testes in which sperm are produced
ovary	production of eggs and sex hormones
scrotum	pouches of skin outside the abdomen that enclose the testes; provide a cooler environment for sperm development
epididymis	sperm storage and maturation
uterus	site of embryo implantation and fetal development
male accessory glands	production of lubricating mucus and secretion of seminal fluid; provides a fluid transportation medium for sperm
vas deferens	transportation of sperm
vagina	muscular tube that receives sperm; also serves as the birth canal
oviduct (fallopian tube)	location where fertilization occurs; transports fertilized egg towards uterus
urethra	muscular tube that carries urine from the urinary bladder to the outside; in males, also conveys semen during ejaculation
penis	delivers sperm to female reproductive tract during sexual intercourse

2.   Refer to the labeled photograph in Figure 15-8 below.

urethra   urinary   testis   vas   umbilical   pelvic   inguinal
            bladder   deferens   artery   bones   canal

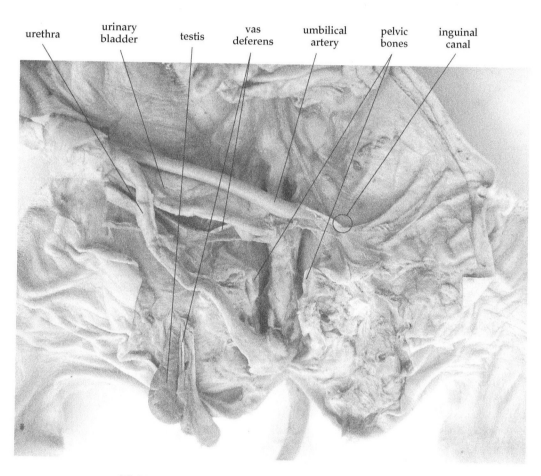

FIGURE 15-8.   Reproductive System of the Male Pig

3.   Refer to the labeled photograph in Figure 15-9 below.

umbilical   urinary   uterine
arteries   bladder   horns   vagina   ovary

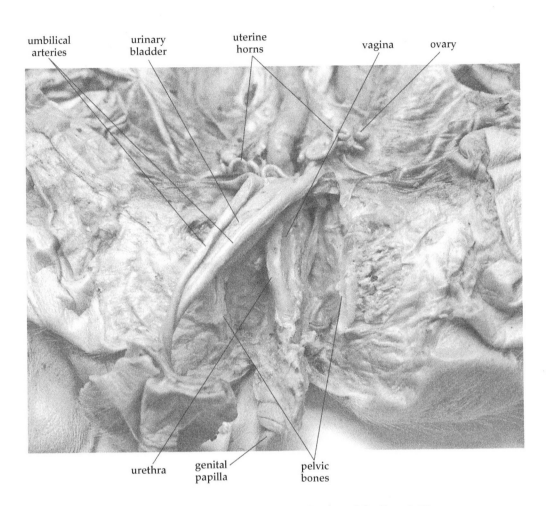

urethra   genital   pelvic
papilla   bones

FIGURE 15-9.   Reproductive System of the Female Pig

4.   After ovulation, the ruptured follicle is converted into a temporary endocrine organ, the corpus luteum. The corpus luteum produces the hormone progesterone.

5.   In addition to the production of gametes, the ovaries also produce female sex hormones. The hormone estrogen is needed for multiple functions around the body that are unrelated to the reproductive system.

6. Yes, since optimal sperm production requires a slightly lower temperature, this could be a problem. Tight pants cause the testes within the scrotum to be held closer to the body, which is several degrees higher in temperature than the scrotum would normally maintain. As an interesting note, research has shown lower sperm counts in professional cyclists. Sitting on a bicycle seat positions the testes closer to the body.

7. In males, the urethra carries both semen and urine. In females, the urethra carries only urine.

8. A corpus luteum forms from a follicle after ovulation. Ovulation does not occur prior to puberty.

Exercise 16 — Mitosis and Asexual Reproduction

1. E — nuclear membrane

2. I — equator

3. C — centromere

4. K — daughter cells

5. A — diploid

6. D — spindle fibers

7. H — cleavage furrow

8. B — sister chromatid

9. G — cytokinesis

10. L — chromosome

11. False

12. True

13. True

14. True

15.   Stages of mitosis for parent cell with four chromosomes:

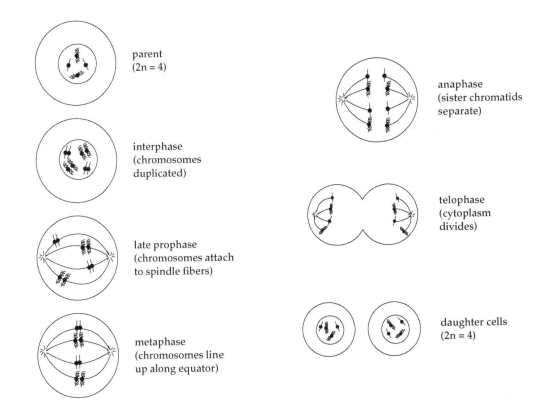

parent
(2n = 4)

interphase
(chromosomes
duplicated)

late prophase
(chromosomes attach
to spindle fibers)

metaphase
(chromosomes line
up along equator)

anaphase
(sister chromatids
separate)

telophase
(cytoplasm
divides)

daughter cells
(2n = 4)

16.   The plan will not be effective. The fishermen are actually doubling the number of sea stars. Sea stars are capable of regeneration, which is an example of asexual cell division. Each half of the sea star can replace lost body tissues and develop into a new, fully functional adult.

17.   A — Skin cells are diploid and have a full set of chromosomes necessary to provide the genetic information to make a new body. The other two options are not viable since sperm cells are haploid and red blood cells have no nucleus.

Exercise 17 — Connecting Meiosis and Genetics

1.   True/False

    F   Sperm cells have the same number of chromosomes as the man who produced them.

    T   A zygote is a fertilized egg.

F   Crossing-over occurs in both mitosis and meiosis.

F   Gametes are genetically identical to each other.

F   The letters AA are used to indicate a heterozygous phenotype.

T   Sister chromatids are identical before crossing-over occurs.

F   Sister chromatids are held together by a structure called synapsis.

T   Daughter cells produced by meiosis are haploid.

T   When the chromosomes line up during metaphase I of meiosis, the equator separates the maternal and paternal members of each pair.

F   The daughter cells produced by meiosis have both a maternal and paternal chromosome from each homologous pair.

2.   Yes. Due to separation of homologous chromosomes in meiosis, each gamete that's produced will have a random mixture of maternal and paternal chromosomes. Recall: Half your chromosomes were inherited from your mother, and (if you're male), you'll pass these on to the next generation in your sperm.

3.   Chromosomes drawn as straight lines represent a single copy of the DNA strand. Before cell division of any type, DNA replication occurs. Each chromosome will then have two copies attached together with a centromere. These appear in an "X" pattern when viewed under a microscope.

4.   The father   Aa

        The mother   Aa

        The child   aa

        Probability of an albino child   25%

        Probability of normally pigmented child   75%

        Explanation: A child inherits **one** allele from each parent; therefore, each parent must have one little "a."

5.   Mother   Tt

        Father   Tt

        First child tt

Explanation: A child inherits **one** allele from each parent; therefore, each parent must have one little "t." The parents could not, however, be homozygous recessive, since people with Tay-Sachs disease die in early childhood.

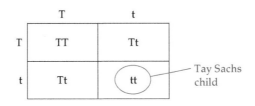

6.   Ghandi   Cc

     Sabrina   Cc

     Snowflake   cc

     Probability of heterozygous cub   50%

7.   Ghandi   Cc

     White tiger at zoo   cc

     Probability of another white cub   50%

Exercise 18 — Human Genetics

1.   The woman   tt

     Her husband   Tt

     The two sons   Tt

     The daughter   tt

     Grandparents   Tt and T?

2.   Your genotype   Dd

     Your husband   DD

     Your sister   dd

     Probability of Tay-Sachs child   0%

3. Fred $X^B Y^o$

    Ginger $X^B X^b$

    David $X^b Y^o$

    Takiyah $X^B X^?$ but probably $X^B X^B$

    Kelly $X^B X^b$

    Kevin $X^b Y^o$

    Takiyah's five sons $X^B Y^o$

    Probability of color-blind son   25%

    Probability of color-blind daughter   0%

4. Ralph $X^H Y^o$

    Ralph's brothers $X^h Y^o$

    Ralph's sister $X^h X^h$

    Ralph's mother $X^H X^h$

    Ralph's father $X^h Y^o$

    Since Ralph has normal vision, he must have inherited a big "H" from his mother. Since Ralph has a color-blind sister, however, his mother must be heterozygous ($X^H X^h$). His father must be color-blind (in order to produce a daughter with the genotype $X^h X^h$).

5. Pedigree showing albino individuals:

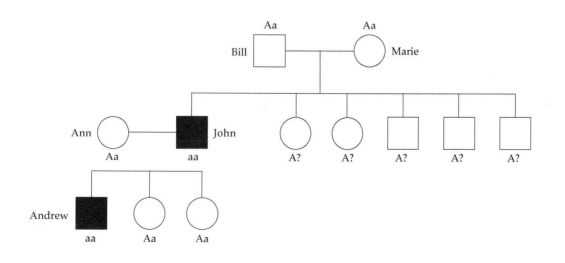

Albinism is inherited through a recessive allele. Neither Bill nor Marie are albinos, yet they had an albino son, John. In a dominant/recessive trait, this is only possible if both parents carried (but didn't express) a recessive allele for the trait. An albino individual would be homozygous recessive, and it wouldn't be possible for two albino parents to have a child with normal skin color, since neither parent possesses an allele for normal pigmentation to pass on to their offspring.

6. Based on the results of a blood test, it isn't possible to conclude that Neil and Alice have the wrong baby. Since Neil has type A blood, his genotype can be either $I^A I^A$ or $I^A i$.

Alice has type B blood, which means her genotype is either $I^B I^B$ or $I^B i$.

The baby has type O blood, which has only one possible genotype, ii. Also, you know that a baby has to inherit one allele from each parent. If Neil's genotype was $I^A i$ and Alices's genotype was $I^B i$, the baby could have inherited an "i" from each parent and have type O blood.

To be completely sure that the baby is theirs, DNA testing would be required.

Exercise 19 — Introduction to Molecular Genetics

1. A — DNA

2. C — both DNA and RNA

3. A — DNA

4. A — DNA

5. D — neither DNA nor RNA

6. B — RNA

7. C — both DNA and RNA

8. B — RNA

9. A — DNA

10. C — both DNA and RNA

11. Yes, your friend has misunderstood. You wouldn't be able to tell this from circulating red blood cells, since they contain no nucleus and, therefore, no DNA.

12. 6

13. C A A

14. G U U

15. Amino acid #1 — isoleucine

    Amino acid #2 — glycine

    Amino acid #3 — leucine

    Amino acid #4 — serine

16. No, because there is more than one codon that codes the same amino acid. In this particular case, both CCT and CCC code for glycine. Other mutations could cause the substitution of an incorrect amino acid in the protein, but not this one.

Exercise 20 — Biotechnology: DNA Analysis

1. Noncoding DNA is that portion of the DNA molecule that has no known function and doesn't code for the synthesis of proteins. Although the functions of these areas are not yet clearly defined, scientists suspect that they play an important role in RNA synthesis and other cellular functions.

2. The pattern of repeating noncoding DNA sequences differs from person to person. This gives each person a unique DNA profile through the process of RFLP analysis.

3. No child is genetically identical to either of the parents. Through the process of meiosis and fertilization, each child inherits half the genes from each parent.

   By analyzing the child's STR profile and comparing it with that of both parents, it's possible to determine which bands were not inherited from the mother, and therefore conclude that they must have been inherited from the father.

   In blood sample comparisons, however, the entire STR profile must be a match between the two samples to show that both the blood samples came from the same person.

4. The deer meat could be tested by STR analysis. If the deer's STR profile matched that of the skin or internal organs found inside the park, it would demonstrate conclusively that the confiscated deer was the individual killed within the park boundaries.

5. The blood found on the dog's collar matches the owner's blood. An analysis of the STR profile of the child's blood does not match that of the sample from the dog's collar. This evidence substantiates the owner's claim that his dog was not responsible for the attack.

Exercise 21 — Using Biotechnology to Assess Ecosystem Damage

1. Product A was most toxic at full strength because the bacteria only glowed at an intensity level of one.

    Product A was determined to be harmless at 1:100 because this was the first dilution at which bacteria glowed at an intensity level of four.

2. There was no dilution level at which Product B was harmless. Even at the lowest dilution (1:10,000), the bacteria only glowed at an intensity level of two.

3. Product A is least harmful to the environment because it is less toxic at lower dilution levels. It can be safely disposed of by diluting with water in the ratio of 1:100

Photo/Illustration Credits

CHAPTER 2

Page 32: Figure 2-4/Celestron International

CHAPTER 3

Page 47: Figure 3-1/BioCam Communications

Page 48: Figure 3-2 a and 3-2 b/BioCam Communications

Page 61: Figure 3-7/Dennis Kunkel Microscopy, Inc.

CHAPTER 6

Page 105: Figure 6-1/Meg Caldwell Ryan

CHAPTER 8

Page 136: Figure 8-1/Photo Researchers, Inc.

Page 139: Figure 8-3/BioCam Communications

Page 147: Figure 8-6/Photo Researchers, Inc.

CHAPTER 12

Page 217: Figure 12-3/Education Resources

CHAPTER 13

Page 233: Figure 13-1/Education Resources

Page 234: Figure 13-2/Education Resources

Page 235: Figure 13-3/Education Resources

Page 238: Figure 13-5/Education Resources

Page 242: Figure 13-7/Education Resources

CHAPTER 14

Page 255: Figure 14-2/Education Resources

Page 259: Figure 14-4/Education Resources

Page 264: Figure 14-5/Education Resources

CHAPTER 15

Page 279: Figure 15-2/Meg Caldwell Ryan

Page 283: Figure 15-4/Meg Caldwell Ryan

Page 290: Figure 15-7/Martini Anatomy and Physiology, 5th edition

Page 294: Figure 15-8/Meg Caldwell Ryan

Page 295: Figure 15-9/Meg Caldwell Ryan

CHAPTER 16

Page 306: Figure 16-4/BioCam Communications

CHAPTER 20

Page 396: Figure 20-4/Orchid Cellmark Diagnostics

Page 404: Figure 20-8/Michael Garvey

Page 408: Figure 20-9/Michael Garvey

Page 409: Figure 20-10/Michael Garvey

Page 410: Figure 20-11/Michael Garvey

Page 411: Figure 20-12/Michael Garvey

Page 412: Figure 20-13/Michael Garvey

Page 418: Figure 20-14/Michael Garvey

Page 419: Figure 20-15/Michael Garvey

Page 420: Figure 20-16/Michael Garvey